EVERYDAY GEOGRAPH

Humph

HODDER AND STOUGHTON
LONDON   SYDNEY   AUCKLAND   TORONTO

Notes for the Teacher

This book is a resource for students wanting to develop their awareness of the key geographical locations that are basic general knowledge for a citizen in the UK. The book is geared particularly to the outline of the AEB's Basic Test in Geography and covers all the elements included in that outline, and each place mentioned in it. This book is also very relevant to the AEB Basic Test (Special) in Geography for Tourism and Leisure. Almost every element of that outline is covered here as well.

The way the book is used is likely to vary according to the group involved. Teachers will no doubt develop the style and the scope of the tasks according to their normal teaching approach. The book is not seen as the sole basis of a two-year Geography course.

Some aspects of the AEB test outline are very clearly defined, but others cannot be laid down so precisely. This is particularly true of the background details required for the UK and world locations. The aim here has been to encourage the student to seek out some key points about each of the locations. The information presented is an indication of the type of detail to seek out but it is not meant to be, and could not manage to be, a compendium of all the points that might arise in an AEB test.

The outline of the AEB test does not lead to detailed study of geographical areas but is intended to ensure some working knowledge of the landmarks of the United Kingdom, the nearer parts of Europe, and the rest of the world in a limited way. Each section of the book has been prepared to suit the needs of all students. Generally the later questions in a series involve the details that may not be essential to a student who is aiming at achieving a pass grade only. These questions will often involve investigations based on other standard reference material, particularly atlases. Not all the answers are included in this book.

The arrangement of the sections of the book follows the outline of the AEB Basic Geography test. This begins with a section on skills, continues with a section on locational knowledge, and has a third section on problem work. The AEB also specifies that centres entering candidates should ensure that they have undertaken certain aspects of practical work, in particular a study of another town. The book contains two outlines of the type of study that can be undertaken in this way (both being based on real papers used in GCSE examinations). These examples assume that the candidates travel to a town they do not know well. On arrival they are given the question paper and street plan, and spend two hours around the town collecting the information they require. Then, in the afternoon, they have two hours to write their answers under examination conditions, using the information gathered in the morning. The candidates may need to travel 50 miles or so to the town concerned. It is hoped that the examples in this book will help teachers to develop similar work based on towns in their own area.

Note: Imperial and metric measurements are used in the book to encourage familiarity with both systems.

British Library Cataloguing in Publication Data

Dobinson, Humphrey M. (Humphrey Mark), 1938–
    Everyday geography
    1. Geography
    I. Title
    910

ISBN 0 340 40286 5

First published 1989

Copyright © 1989 Humphrey M. Dobinson MBE

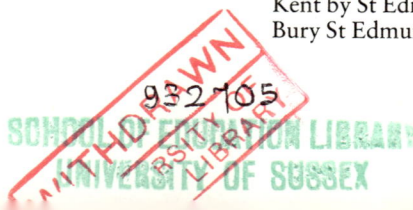

Typeset by Rowland Phototypesetting Ltd,
Bury St Edmunds, Suffolk
Printed in Great Britain for Hodder and Stoughton Educational, a division of Hodder and Stoughton Ltd, Mill Rd, Dunton Green, Sevenoaks, Kent by St Edmundsbury Press Ltd, Bury St Edmunds, Suffolk

Contents

Town Plans

Chepstow is an old town not far from Bristol. It is on a hill near the place where the River Wye joins the River Severn. A castle was built there in 1067 by one of the Norman barons helping William the Conqueror to take control of England. The town developed later as a market and as a port. For many years, there was an important industry building wooden ships.

On these pages, the town of Chepstow is shown:

- as it is seen in the main streets (Fig. 1.1)
- as it is seen from the air (Fig. 1.2)
- on a tourist map of the district (Fig. 1.3)
- on a town plan (Fig. 1.4)

The Ordnance Survey map of the area is on the back cover of this book.

Every place looks different on a map, and not the same as it seems when you are there. The reason is that maps are drawn according to certain rules. One of these rules is that every place is shown as you would see it from directly above. Another reason is that maps show buildings, trees, and other landmarks as symbols, not as pictures. A third rule is that maps only show things that are always there, and not things that come and go.

Fig 1.2 ● Aerial view of Chepstow

Fig 1.3 ● Map of Chepstow district. (*Wye Valley AONB Joint Advisory Committee 1984. Based on the Ordnance Survey map with the permission of the Controller of H.M. Stationery Office, Crown copyright reserved*)

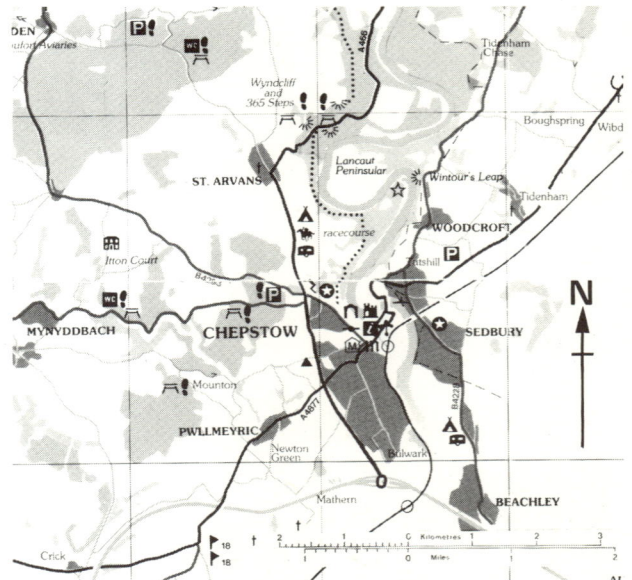

Fig 1.1 ● Chepstow street scene

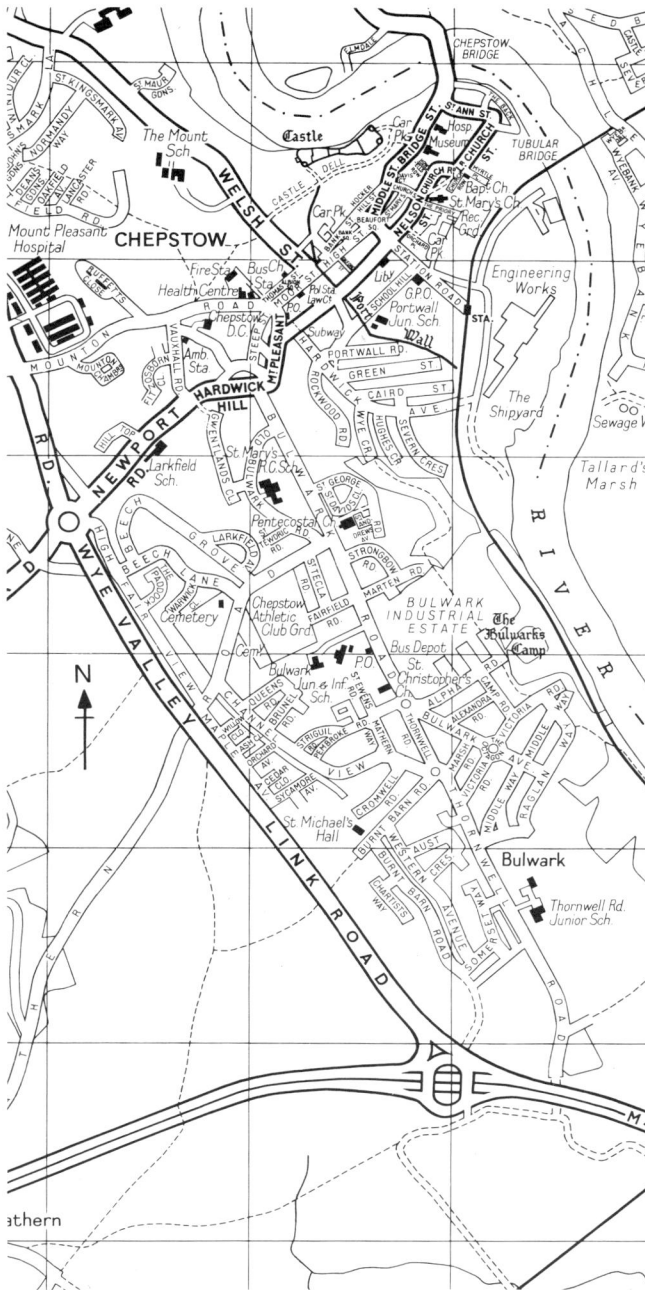

Fig 1.4 ● Part of town plan of Chepstow

1. How many reasons can you suggest why a photograph taken from the air does not show streets exactly as they are on a town plan? (You may want to use the words *distance*, *perspective*, *hidden*, *temporary* in your answers.)

2. (a) Describe your route from the Wye Valley Link Road – motorway (M4) junction to the castle.

(b) What details about that journey can you find in the aerial photograph that are not shown on the town plan?

(c) What details are shown on the town plan that are not clear in the aerial photograph?

3. Describe a route walking from the Post Office in Bulwark Road to Mount Pleasant Hospital.

4. One of the photographs shows the castle. What can you find out from the photograph that is not shown on the town plan?

5. Figure 1.1 shows part of the main street. If you needed directions to reach the bus station, why might the town plan be more helpful than the photograph?

6. What ten attractions are there for tourists in Chepstow and the area just around the town?

7. From which angle were the photographs of the bridge and the castle taken? What are the landmarks on the photographs that you can find on the map?

8. What else can you find out about the river and the buildings that is not shown on the map?

9. Why are maps usually more useful than a set of photographs?

Fig 1.6 ● Chepstow Castle

Fig 1.5 ● Chepstow road bridge

Grid Squares

Most maps show hundreds, or even thousands, of names or symbols. If you have no idea where Rushcliffe Rise is, it can take a long time to find it if every street is named.

To make it easier to find a particular place, lines are drawn across the map to divide it into squares. Squares are usually given letters or numbers along the bottom of the page, starting from the left hand corner. There are usually numbers up the side of the page, but on some maps it is the other way round. These letters are called the *co-ordinates*, and the list of names is the *gazetteer*.

Each place is then listed in alphabetical order, with the number and letter of the square in which it is shown. When you have found the right square, all you have to do is to check through the places shown in that square to find the name you want.

The map on this page shows the centre of Newbury, a small town in Berkshire.

1. What are the co-ordinates of the square that has been shaded?
2. Through which squares does the railway line run?
3. In which square is the railway station?
4. The photograph shows the Town Hall. In which square is it?
5. In which square is the hospital?
6. Highfield Avenue is in square B4. What street must you go along before you can turn into it?

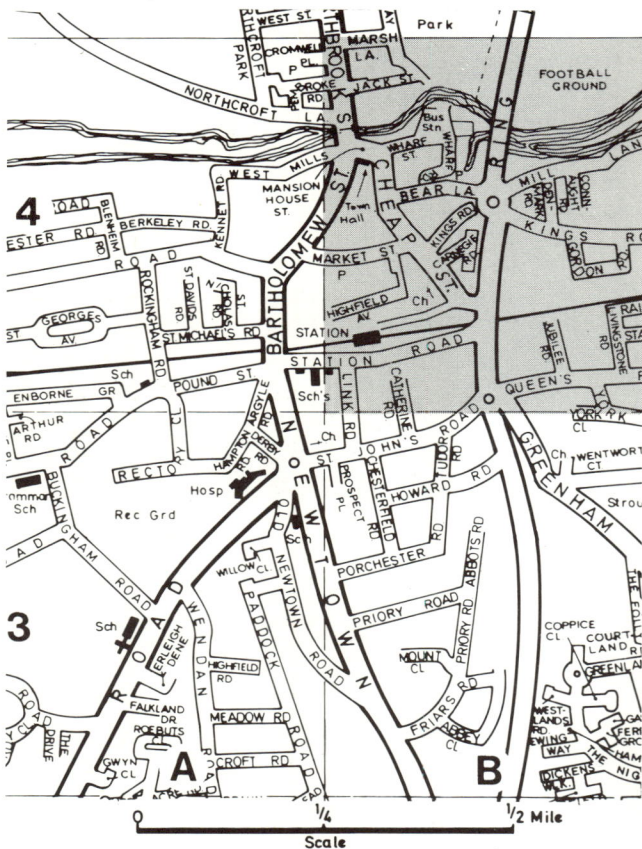

Fig 1.7 ● Part of a map of central Newbury

Fig 1.8 ● Newbury Town Hall

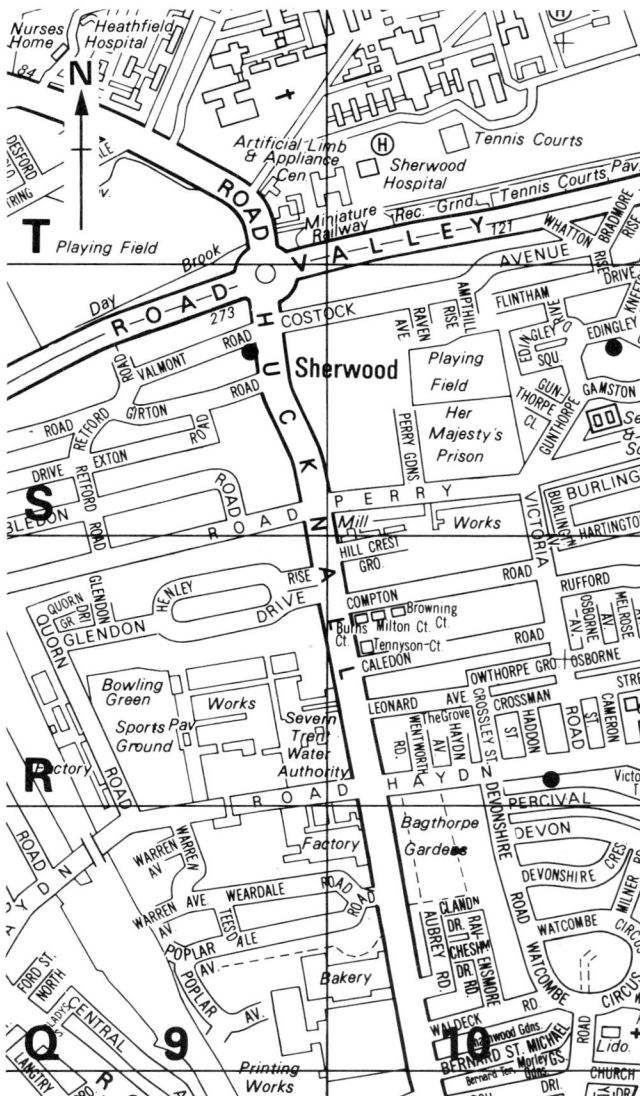

Fig 1.9 ● Part of a street map of Nottingham. (*Reproduced by permission of Geographers' A–Z Company Ltd. Based on the Ordnance Survey map with the permission of the Controller of H.M. Stationery Office, Crown copyright reserved*)

Girton Road	S9
Glade Hill	W10
Gladhill Rd	W11
Gladstone St	O8
Glendon Drive	R9
Goodwood Ave	X13
Grampian Drive	Z9
Gretton Rd	S15
Gunthorpe Close	S10
Haddow St	R10
Hall Clinic	V10
Hallam Rd	Q15
Hallam's Lane	W14
Hampstead Rd	P13
Hardwick Rd	R11
Hartington Rd	S10
Hawthorns	V10
Haydn Ave	R10
Haydn Rd	R10
Hazel Grove	S16
Heathfield Hospital	T9
Heathfield Rd	T7
Heaton Close	P15
Henley Rise	R9
Hexham Gardens	Z8
High Pavement College	U8
High St	W14
Hill Crest Grove	R10

● Part of gazetteer

On this page there is a street plan for part of the city of Nottingham. There is also part of the gazetteer for that plan.

1. Use the gazetteer and the street plan to find these eight streets or places. When you have found each one, write down: (a) the co-ordinates for the square where this street or place can be found; (b) the name of another street that you can turn into from this street or place, as shown on the street plan.

Girton Road
Glendon Drive
Gunthorpe Close
Hartington Road
Haydn Ave
Heathfield Hospital
Hill Crest Grove

2. The following streets are listed on the gazetteer and shown on other parts of the map. Would you look to the north, south, east, or west of the part of the map shown here, to find each one?

Hall Clinic
Hazel Grove
Heathfield Road
Glade Hill
Grampian Drive
Hardwick Road

Road and Rail Maps

Road Maps

There are many different sorts of maps because people have many different needs. It is a good idea to make sure you have the right map for the information you want.

A large scale map is one that shows each place at a larger size than on a small scale map. You see more detail on a large scale map, but it does not cover such a big area.

Each map aims to show clearly the information the reader wants to know and leaves out other information. A map that showed everything would be so full that it could not be read. A Throughway Town Plan map leaves out the names of the small side streets you pass as you go through. The Complete UK Towns and Villages map shows where all the places are, but does not include roads, rivers, or hills, while the Leisure Map highlights interesting things to see, and how to get there.

Look at the following list of customers in a map shop. Decide who is likely to want each of the maps listed here:

Types of map:

- Throughway town plans
- Leisure Map (Tourist Board)
- Atlas of roads of Britain
- Town Plan (large scale)
- Town and district
- UK main routes map
- Principal coach and railway stations
- UK physical features (hills and rivers)
- UK complete towns and villages
- Airports and seaports

Customers in map shop:

- Long distance lorry driver
- Commercial representative
- Van delivery driver
- Visitor on holiday
- New resident in a town
- Taxi driver
- Transport manager, export department
- Scenic photographer
- Student
- Journalist or news editor

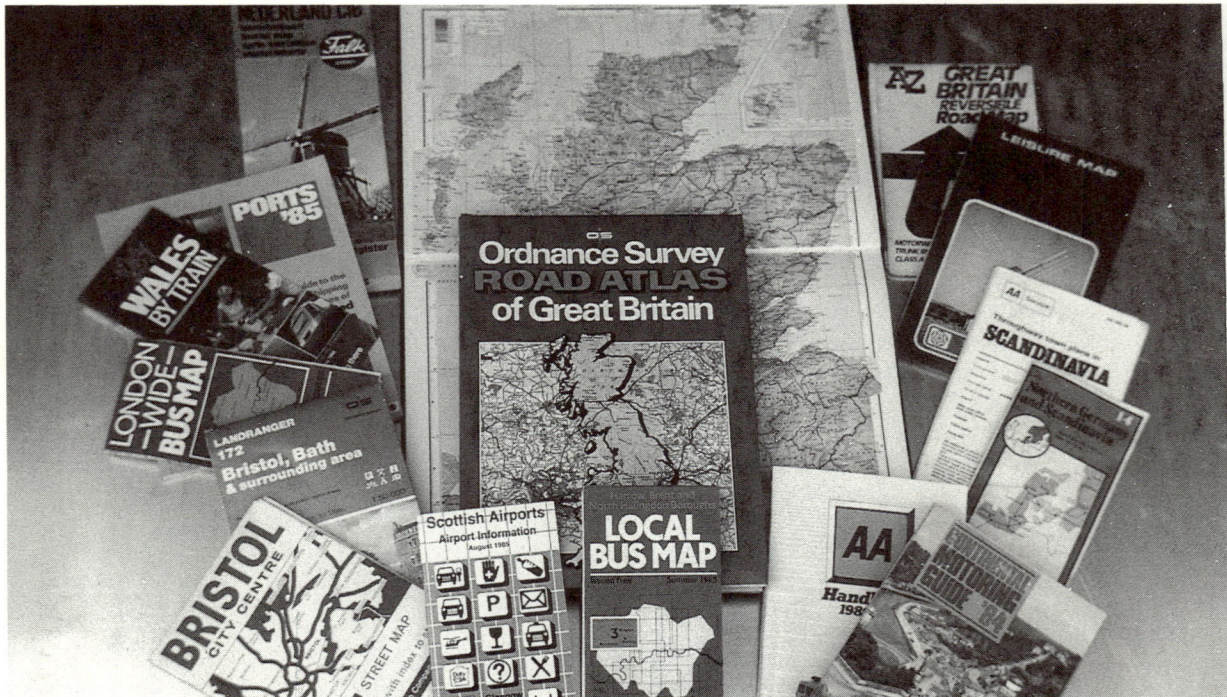

Rail Maps

Maps that show routes by train are even simpler than road maps. As a passenger, you do not need to know where the line curves, or where the points or signals are. These maps therefore often show the routes between towns as being straight lines.

For making a journey across the part of Europe shown on this map, there is a choice of routes you could take.

1. Which is the shortest route to go from Calais to Hanover?
2. Which route from Calais to Munich has the smallest number of junctions, shown on corners on the topological map?
3. Which route from Calais to Milan crosses the smallest number of frontiers where you might need to show passports?
4. Which route from Calais to Bremen would be best if you want to call on a friend in Brussels?
5. Which route from Calais to Venice would be best if you wanted to see as much as possible of Switzerland on the journey?

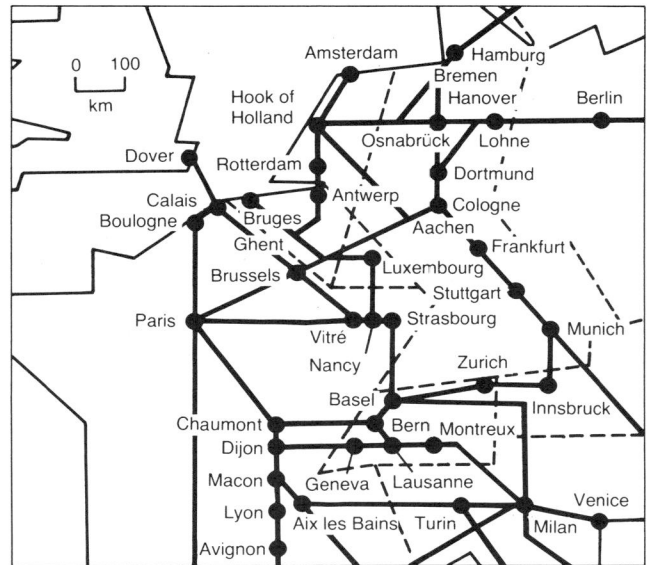

Fig 1.10 ● Rail map for part of Europe

Fig 1.11 ● Area maps of Europe

Choosing the Right Area Maps

The outline map of Europe shows the number of the map for each area featured in a European Atlas. Use an atlas to find the places mentioned.

1. Which maps would you need if you were travelling from Harwich to Paris?
2. Which five maps would you need for a holiday starting in France at Cherbourg and visiting Bordeaux, Avignon, Lyons, and coming back through Luxembourg and Brussels?
3. Which other map also shows Luxembourg?
4. What is different about maps 2 and 3? Why would you have to remember this difference while using them to plan what you do each day?
5. Which maps would you need if you flew to Nice and then visited Corsica, Rome and Florence?
6. Which maps would you need if you had to drive a lorry from London to Hamburg?
7. Which map would you need if you were going to Brussels?
8. On which map is Gothenburg (Sweden)?
9. If you were just visiting the area around Oulu in Finland, you could choose either of two maps. Which two could they be?
10. Which map is entirely repeated on parts of three other maps? Why is it available separately?

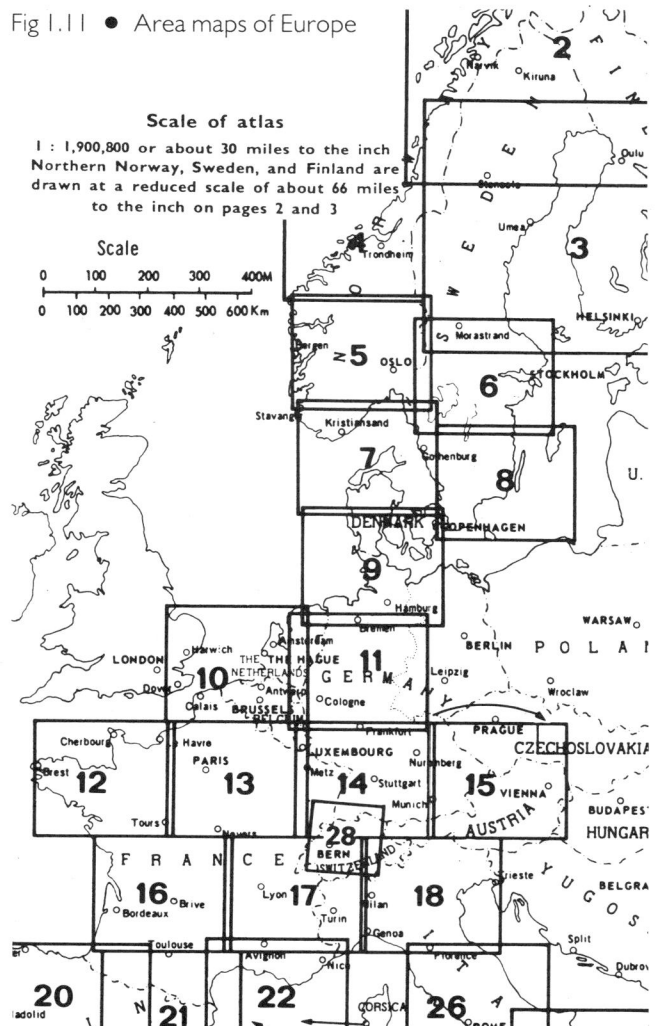

Topological Maps

Some maps show special routes e.g. those taken by public service vehicles. The places served by the routes are shown at convenient points, but these points are not marked in their true locations. Maps that are laid out in this way are called 'topological' maps.

On this page there is a topological map of the principal routes run by British Rail. Use it to answer these questions:

1. What stations are shown on the route from Birmingham New Street to Carlisle?
2. Find the places you have just listed on an ordinary map of England. Put a ruler on the map to show a straight line from Birmingham to Carlisle. Say which of the places named would lie to the west of the straight line.
3. The distance from Lancaster to Carlisle is really much more than the distance from Preston to Crewe. Why does the British Rail map seem to show these distances incorrectly?
4. The map of London is greatly enlarged on the British Rail map. Why was it not good enough just to show London as a big circle on roughly the same scale as the rest?
5. What route would you take to go from Grimsby to Harrogate?
6. What route would you take to go from Cardiff to Edinburgh?
7. What route would you take to go from Bournemouth to Dover?
8. What route would you take to go from Exeter to King's Lynn (using London Transport as necessary)?

Fig 1.12 ● British Rail main routes

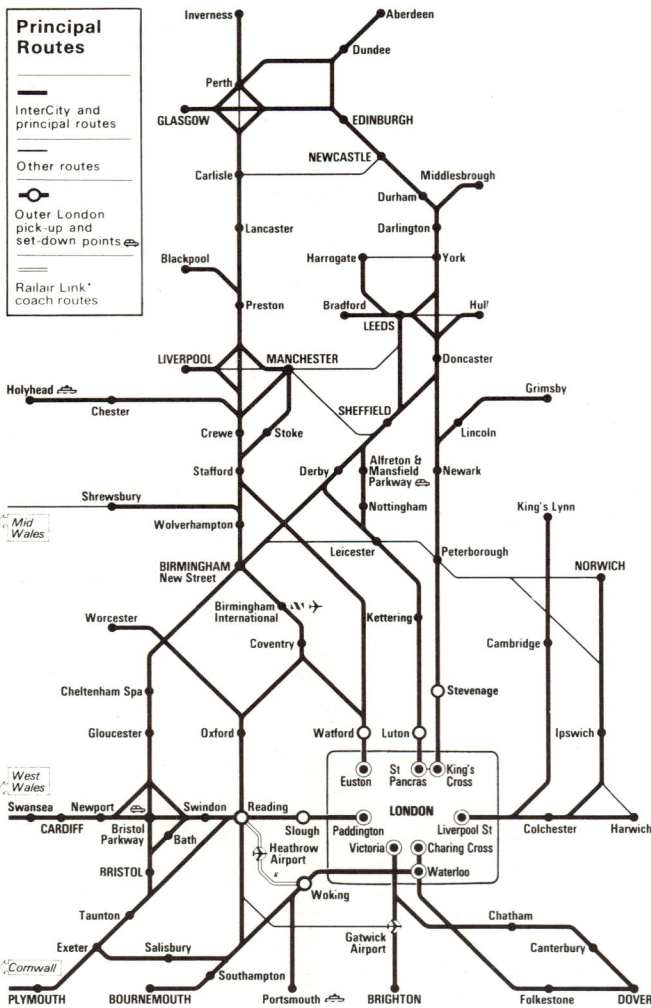

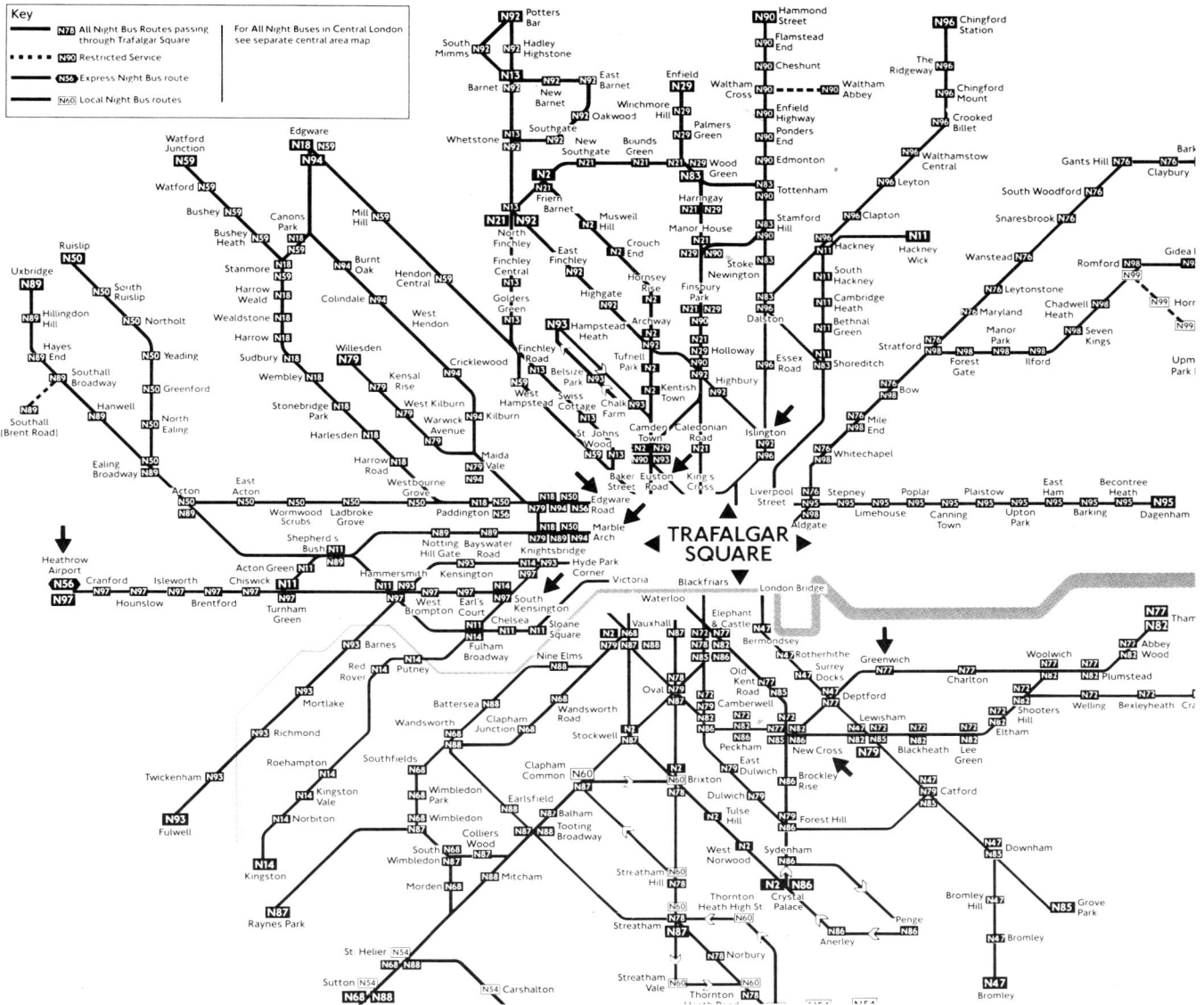

Fig 1.13 ● Part of the 'All Night' London bus map

The topological map above shows the bus routes running all night in the centre of London. When you find the same number in the circles at each end of your journey you know which bus you need.

1. Which four buses could you use to go from Edgware Road to Marble Arch?
2. Which bus will take you from Heathrow Airport to South Kensington?
3. What named place would you pass on a N77 bus from Greenwich to New Cross?
4. What eight named places can you travel to on a N2 bus going north out of Euston Road?
5. How far can you go on the N96 from Islington?
6. Why was the map not drawn to scale?

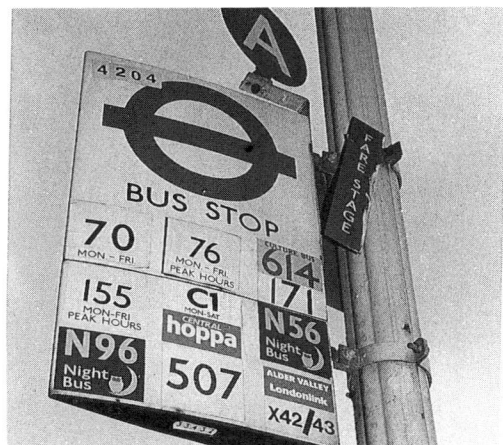

Fig 1.14 ● Bus stop sign including 'All Night' routes

Floor Plans

A special sort of map is the one that shows you the way around a large school or college, or a big department store, or an exhibition or show. There may be a red star on it marked 'You are here'. Then you have to decide which of the floors you want to reach and which way you have to turn.

You have arrived on the car deck of a ferry (see Figure 1.15). You have found the lift, and arrived near the stern of Deck 5 among a crowd of people, all speaking foreign languages. Say how you would reach each of these places from where you are now. You can either give directions by saying 'left' and 'right', 'back' and 'front', or you can use the ship's words 'port' and 'starboard', 'stern' and 'bow'.

1. To go from the stern lift to the shopping arcade. (Which direction? How far? Which side?)
2. To go from the shopping arcade to the cafeteria room. (Which deck? How to get there? Which side?)
3. To take a child from the shopping arcade to the play ship.
4. To go from the play ship to cabin 5800.
5. To go from cabin 5800 to the ship's cinemas.
6. To go from the cinemas to the pool.

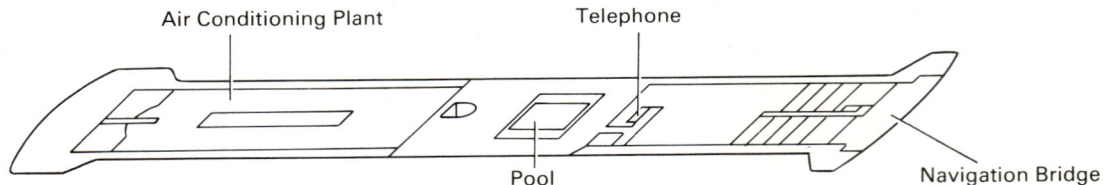

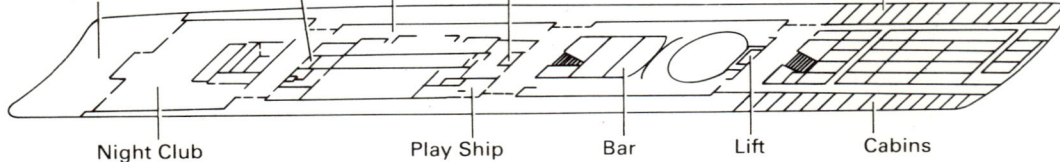

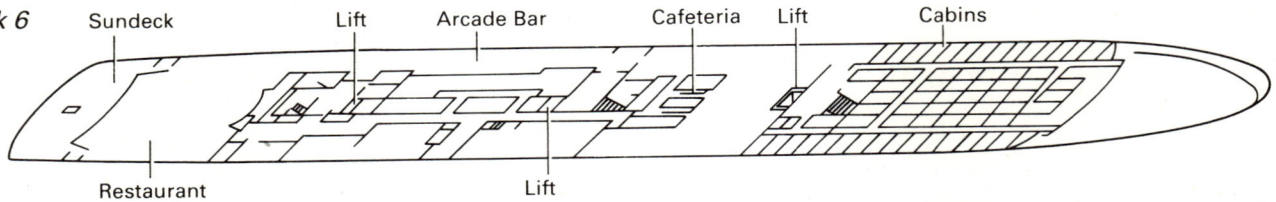

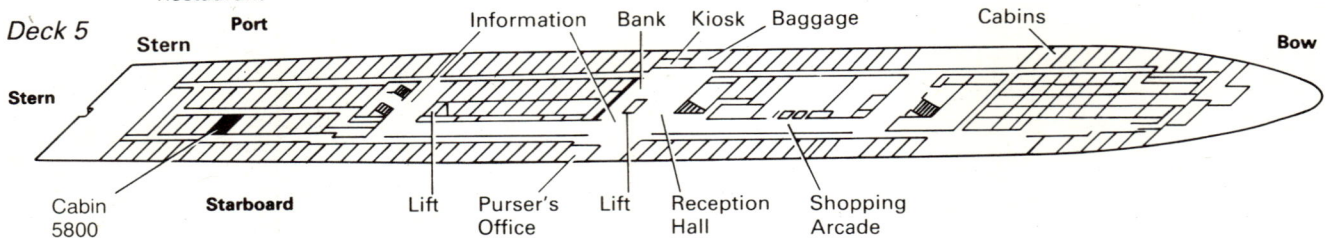

Fig 1.15 ● Deck floor plans of a ferry

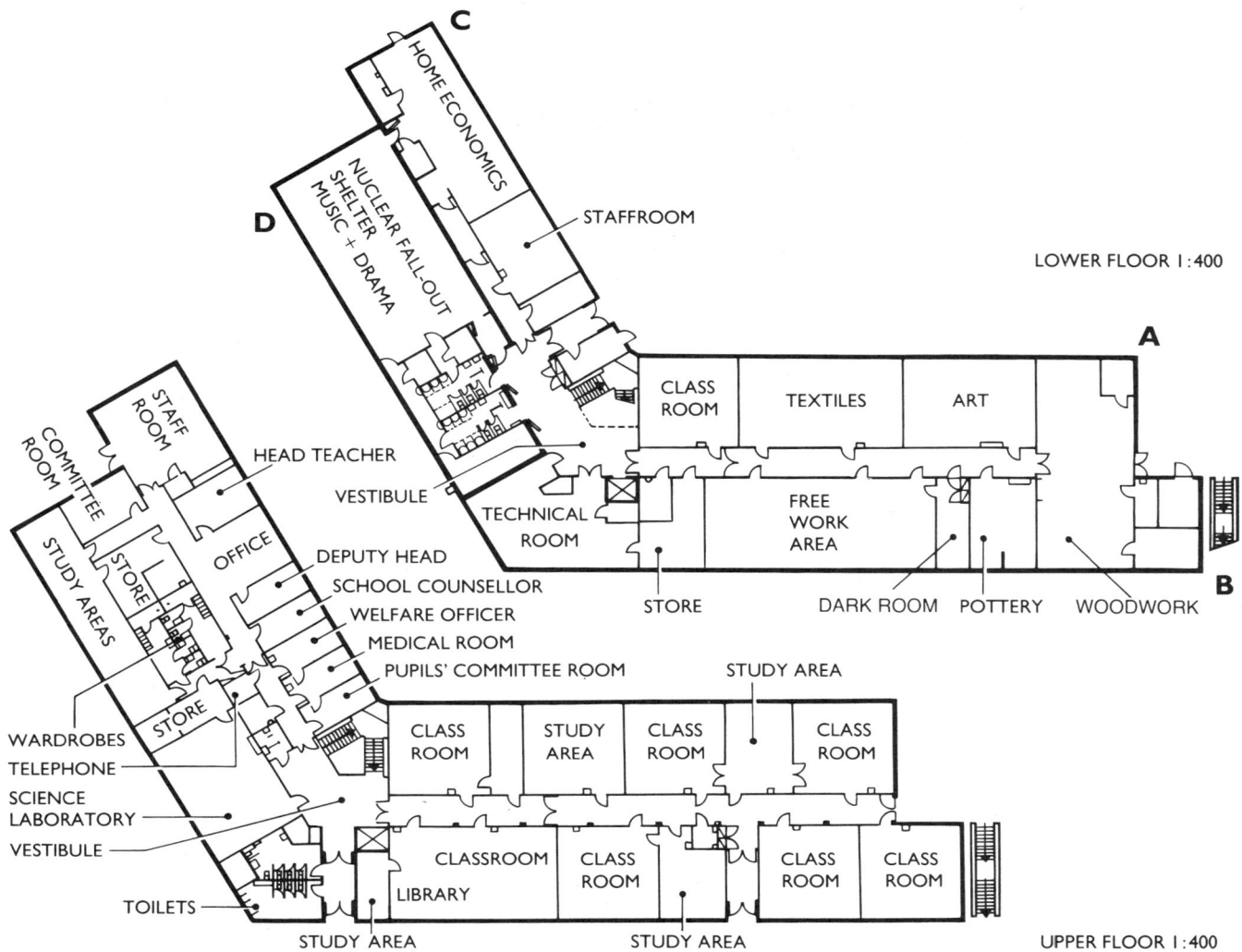

Fig 1.16 ● Floor plans of a secondary school

These are the floor plans for a new secondary school. Would you like all schools to be like this? Can you find your way to all the right rooms straight away?

1. How many sets of doors would you pass from the main entrance on the lower floor to the home economics classroom?
2. Give directions to go from the art room to the nuclear fall-out shelter.
3. Give directions to reach the Head's office from the upper floor main entrance.
4. How would you go from the home economics room to the library?
5. How would you go from the textile room to the pupils' committee room?
6. Shelley was looking for the photographic darkroom but found an upper floor entrance. What had she done wrong?

7. Aaron was looking for the School Counsellor but found the Deputy Head's office. What mistake had he made?
8. From which of the points, A, B, C, or D, was the photograph taken?

Fig 1.17 ● Outside view of the school

Ordnance Survey Maps

Fig 1.18 ● The Severn Bridge

The most detailed ordinary maps on sale to the public are the Ordnance Survey maps. The maps of England, Scotland and Wales are produced by the staff at a government office in Southampton. The information they obtain is published in maps on many different scales. The map that is sold most often is the Landranger map, on a scale of 1 to 50000. A total of 204 different sheets cover the whole of Great Britain. (There are OS offices in Northern Ireland and the Republic of Ireland as well.)

There is so much information on a Landranger map that it has to be printed in several colours. The Ordnance Survey can include a wide range of details since one colour can be printed on top of another. This means that contour lines showing the height of the land can be drawn in orange with no breaks. These can be seen even where woods are shaded green or where built up areas are shaded grey.

Note: To answer the following questions, you will need the standard Ordnance Survey key found on any map in the Landranger series. These questions are based on part of sheet 172, which is printed on the back cover of this book.

Place names and river names (printed in black or blue)

1. Which is the largest town marked on the map?
2. What are the names of the villages at either end of the big bridge?
3. What is the name of the wide river, and where can you find it printed?
4. What is the name of the smaller river that joins the big wide one?
5. What 'lakes' are marked on the map?

6. What are the names of the drainage channels south of Aust?

7. What is the name of the V-shaped area of water in the north-east corner of the map, and what industrial building is beside it?

8. The map shows parts of three counties. How are the borders between these counties shown?

9. In which county is Beachley Point?

10. What are the names of the two farms in square 6090, south-west of Oldbury-on-Severn?

Routes and route numbers (various colours)

1. Which motorway has an interchange with the M5 near Almondsbury? How many levels are there on this interchange?

2. What is the route number of the main road running south from Aust?

3. How can you tell that the B4055 is not such a big road as the one in question 2?

4. What is the route number of the road from Beachley to Sedbury?

5. What does the thick black line that is near the west bank of the River Wye show?

6. Why are there small lines on either side of this thick black line south of Tutshill?

7. What colour is used to show the road from Pilning to Olveston (6087), and what information is given by this colour?

8. What colour is used to show the road from Elberton to Aust, and what information is given by the dotted lines near Red Hill?

9. If you wanted to walk from the church at Northwick (5686) to Aust, what colour and sort of line would you be following on the map?

10. What colour is used to show the streets (not the buildings) in the built up area of Chepstow?

Symbols

1. If you stopped on holiday at Aust, where would you be able to spend the night?

2. How could you be sure that you would be able to draw money from your National Savings Bank book?

3. Why might something on the map make you want to visit Aust Rock?

4. What would you see at Cote Farm (5889), just north of Aust?

5. Where could you spend a night cheaply at Chepstow (5293)?

6. What four antiquities (things from long ago) are shown in or near Chepstow?

7. How many churches are there in Olveston (6087), and how do they differ?

8. What industrial work has been done at Priestpool, near Elberton (5988)?

9. If you had an accident at the corner of the lane near Priestpool, how could you get help quickly?

10. What building could you use as a landmark to describe where you are at this point?

Physical features

1. What colour are the lines that mark the height of the land?

2. How can you tell that the village of Ingst (5887) is built on top of a small hill? How many metres above sea level is its highest point?

3. What does the small patch of green just west of Old Splott Rhine (5787) show?

4. Why would it be unwise to walk along the river bank to the west of the path at Littleton Warth (5890)?

5. What colour is Slimeroad Sand (5692) shown on the map? What does the use of this colour tell you?

6. What dangerous area is shown by the fine black lines east of Sedbury Park School (5593)?

7. Is the road from Sedbury Park School to Offa's Dyke going uphill or downhill?

8. On what type of land is the path at Littleton Warth (5890)?

9. If you stood at Red Hill (5888) and looked north, how far ahead would the ground be flat?

Fig 1.19 ● Aust Post Office and stores

10. The photograph shows Aust Post Office and was taken facing west. What important building is just beyond the picture?

Routes on an OS Map

Using the Ordnance Survey map on the back cover, choose and describe good routes for these people to follow. To describe the routes say:

- the name of the place where the road or path begins
- the type of path, or road, and where to begin on this path or road
- the compass direction at the point where you begin on this path or road
- the name of the place where you turn onto another path or track, and roughly how far this is
- the names of any places you pass on the way
- describe landmarks you would expect to see on the way

The first one is done as an example:

1. A cyclist wants to take a fairly quiet road from Elberton (6088) to Tockington School (6186).

Answer: The cyclist takes the lane that runs south out of Elberton past the public telephone box and the quarry. After about 1.5 km she will reach Olveston. She should keep straight on past the post office, the church with the tower, and the chapel, and then take the left turn where the road forks. She should not take the sharp left turn that runs north-east to Old Down, but take the road that runs south-east uphill to Tockington. After about another kilometre, she will go downhill into Tockington. She should keep to the left past the pub, and leave the church on her right. Just past the next left turn, she will find the school on her left. Total distance: about 3 km.

2. A cyclist wants to take a quiet road from Hazel Farm near Alveston Down (6288) to St Arild's House (6190).

3. A motorist wants to drive from the power station (6094) to Cowhill (6091).

4. A group who have parked their minibus near the public telephone box beside the A466 (5392) want to walk to the edge of the River Severn.

5. A tourist wants to see Chepstow Castle (5394), Offa's Dyke (5593) and Beachley Point (5490).

6. An electricity board engineer has to check the power lines that run from Red Hill (5888) past Ingst (see photograph on this page). Where will she have to go up and down hill when she walks along the route?

7. A lorry driver with a wide load wants to travel on good roads from the station at Severn Beach (5484) to the station in Chepstow (5393).

8. A photographer wants to travel from Aust to take a picture of the Severn Bridge at Littleton Warth (5890).

Fig 1.20 ● Electricity power lines at Red Hill

Route Maps

Ordinary maps do not show exactly how wide a road is, but they use a key to show whether it is a main road or a minor one. Maps often use a different sort of line to show motorways and dual carriageways. Route maps of this sort leave out details such as rivers, forests, some villages and other landmarks. The roads are highlighted and often the road numbers and distances are shown.

When you are planning a journey in an area you do not know well, you may decide that it is better to take a slightly longer route on main roads. That way you avoid coping with hills, sharp corners and traffic delays on a shorter but minor road.

The map here shows some of the camp sites in the south of France and in Spain. Use it to answer these questions:

1. What towns would you pass through on the road from Toulouse to Narbonne?
2. You decide to spend a night at the camp site at Canet-Plage, near Perpignan (square D4). What is the route number of the main road to it from Narbonne, and where would you decide to leave it?
3. You decide to spend the next night at San Pedro Pescador (square D3). Would you choose the inland or the coastal road? Why?
4. You want to go to Ripoli (square C3). What route numbers do you follow from Figueras and how far is it from Figueras to Ripoli?
5. Next you want to visit Quillan (square C4). Where do you cross the frontier?
6. Suggest (a) a fast route (b) a quieter route from Quillan to Toulouse.

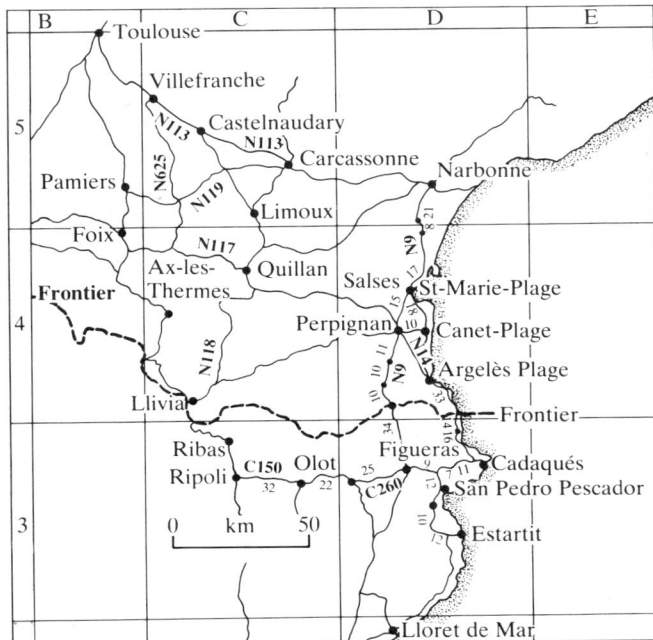

Fig 1.21 ● Some camp sites in South of France and in Spain

Giving Route Directions

Use the maps on this page to describe the routes required, with names, and whether you are turning left or right.

In Uxbridge:

1. To travel from the Underground station to the bus for Oxford, which runs from point G.
2. To find the Underground station when you have arrived on the bus from Hemel Hempstead (at point A).
3. To find the Underground station when you have arrived on the bus from Staines, and got off at point T.
4. To find the post office (which backs onto Oxford Road) from the Underground station.

In Aberdeen:

5. To find the civic centre (on the eastern border of the map) from the post office.

6. To find the railway station if you are arriving on the A956 from Charlestown.
7. To find the P & O Ferries office if you are arriving on the Great Northern Road from Elgin.
8. To find the petrol station in Queens Road if you are in St Machar Drive.
9. To find where the River Dee joins the sea (the estuary) if you are in King Street.
10. To find the Tourist Information office if you are in Holburn Street.

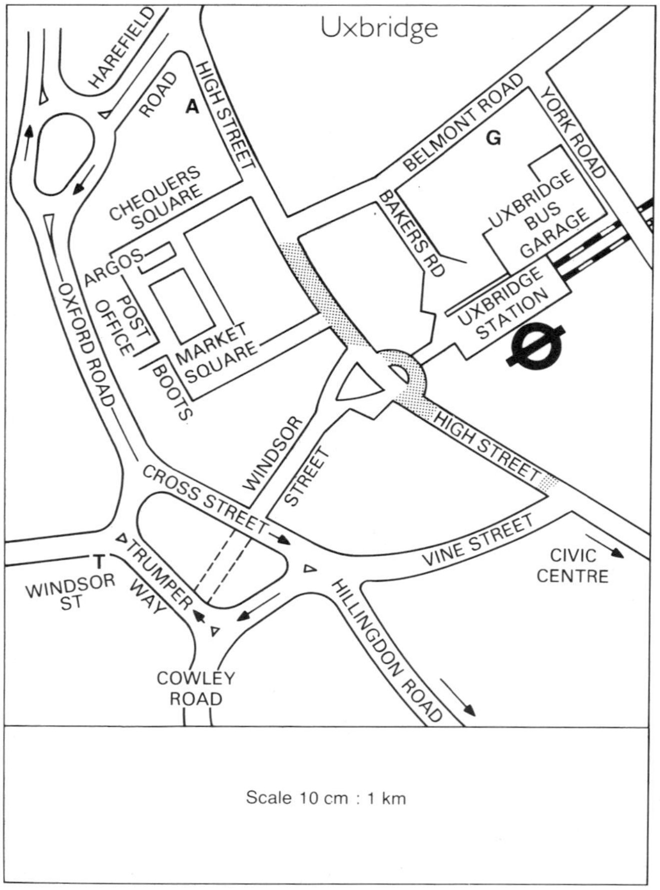

Fig 1.22 ● Map of Uxbridge

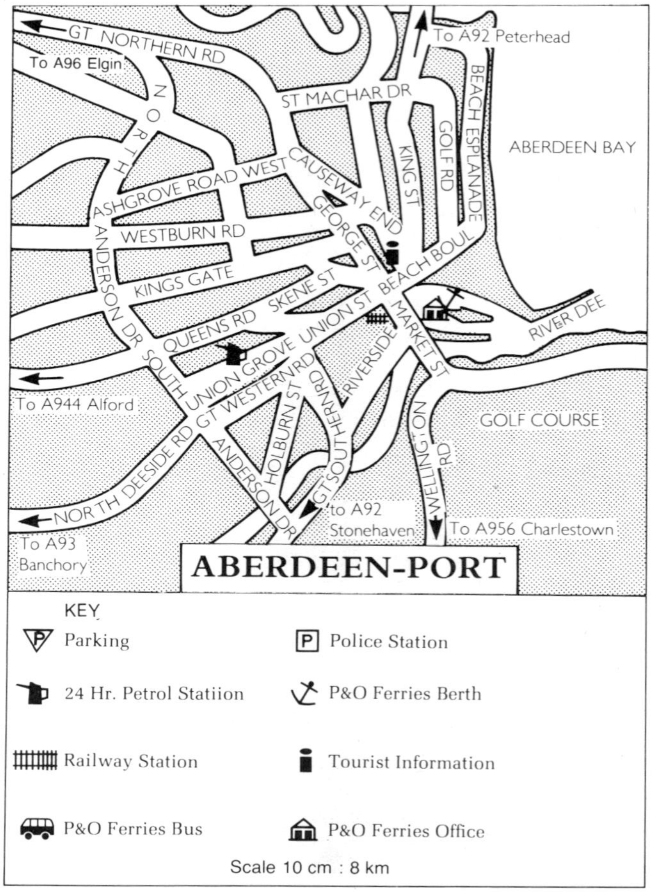

Fig 1.23 ● Map of Aberdeen

Road No.		Miles
	WHITBY	–
A171	Leave by Bagdale	
A169	Junction of roads	2¼
	Turn Left.	
	SLEIGHTS	1½
	PICKERING	17
	MALTON BY-PASS	7
A64	Junction of Roads	
	Turn Right	
	ROUNDABOUT	15½
A1036	Turn Right	
	York City Centre	3
	Leave by Tadcaster Road	
	Junction of Roads	3
A64	Turn Right	
	TADCASTER BY-PASS	10½
	Kiddal Lane End	6½
	LEEDS CITY CENTRE	8¼
	74½ Miles	

Fig 1.24 ● Route Whitby-Leeds

Fig 1.25 ● Part of York City walls and Minster

You can write away for information about the best route between towns you do not know. Here is the information sent by the RAC about the route between Whitby and Leeds through York. When you have read the information they have given, try to describe what the journey would be like. Which bits would you enjoy most? Where might you be held up? Where would you have to be careful not to take a wrong turn? Which part might be the most relaxing?

Here is the information sent by the AA about a route going through the Netherlands and West Germany. The bold numbers on the left show the route number. The other numbers on the left show the total distance travelled on this route (in km), and the distance between each point. The numbers on the right show how far there is still to go.

1. How far is it from Arnhem to Oberhausen?
2. What will you come to after 15.5 km along this route?
3. How far is it from Wesel to the service area at Hunxe?
4. What will you come to at Oberhausen?
5. Is it necessary to use the motorway?
6. Which large river might you see along this route?

(82.5km = 51.5 miles)

This route is by autobahn through the lower Rhine Valley with level and open agricultural country and occasional woods. The motorway can be avoided by following the parallel road (no. 8 in Germany).

Road			Place	Right
E36			*Arnhem	82.5
			roundabout	
			(town 6km) (Plan:C)	
	1.5	1.5	Westervoort/	81
			Giesbeek Junc	
	9	7.5	Didam-Zevenaar Junc	73.5
	15	6	Beek/Babberich Junc	67.5
E36 (A70)	15.5	0.5	Dutch & German Customs of Elten	67
	23	7.5	*Emmerich Junc (town 4km)	59.5
	38	15	*Bocholt/Rees Junc	44.5
	51	13	Bocholt/Wesel Junc	31.5
	65	14	*Wesel/Borken Junc (town 7km) (Iter D188)	17.5
	68.3	3.3	Hünxe Junc (service area)	14.2
	75.1	6.8	Dinslaken Nord Junc	7.4
	78.8	3.7	Dinslaken Sud Junc	3.7
	82.5	3.7	Oberhausen Autobahn Kreuz (motorway X-rds) (Oberhausen 7km; *Essen 20km)	

Fig 1.26 ● Route B17 Arnhem-Oberhausen

Past the Ticket Gate

Thorpe Park, Surrey

There is a lot of fun to be had at Thorpe Park. You don't want to waste a minute! Worst of all is when you spend ages going round and round looking for the wind surfing school and all you can find is the Saxon longhall! With the map that they give you, can you find the things you want straight away?

1. What would you pass between the Admissions building and Mountbatten pavilion?

2. What would you pass between the Children's Rides and the Radio Controlled Boats?

3. Look for the arrow which points north. What would you see next, if you walked north (with your back to the sun) from the Go-Kart track?

4. What would you see next if you were walking west (into the afternoon sun) from the public board sailing school?

5. What would you see next if you turned left on your way back from the Viking longship?

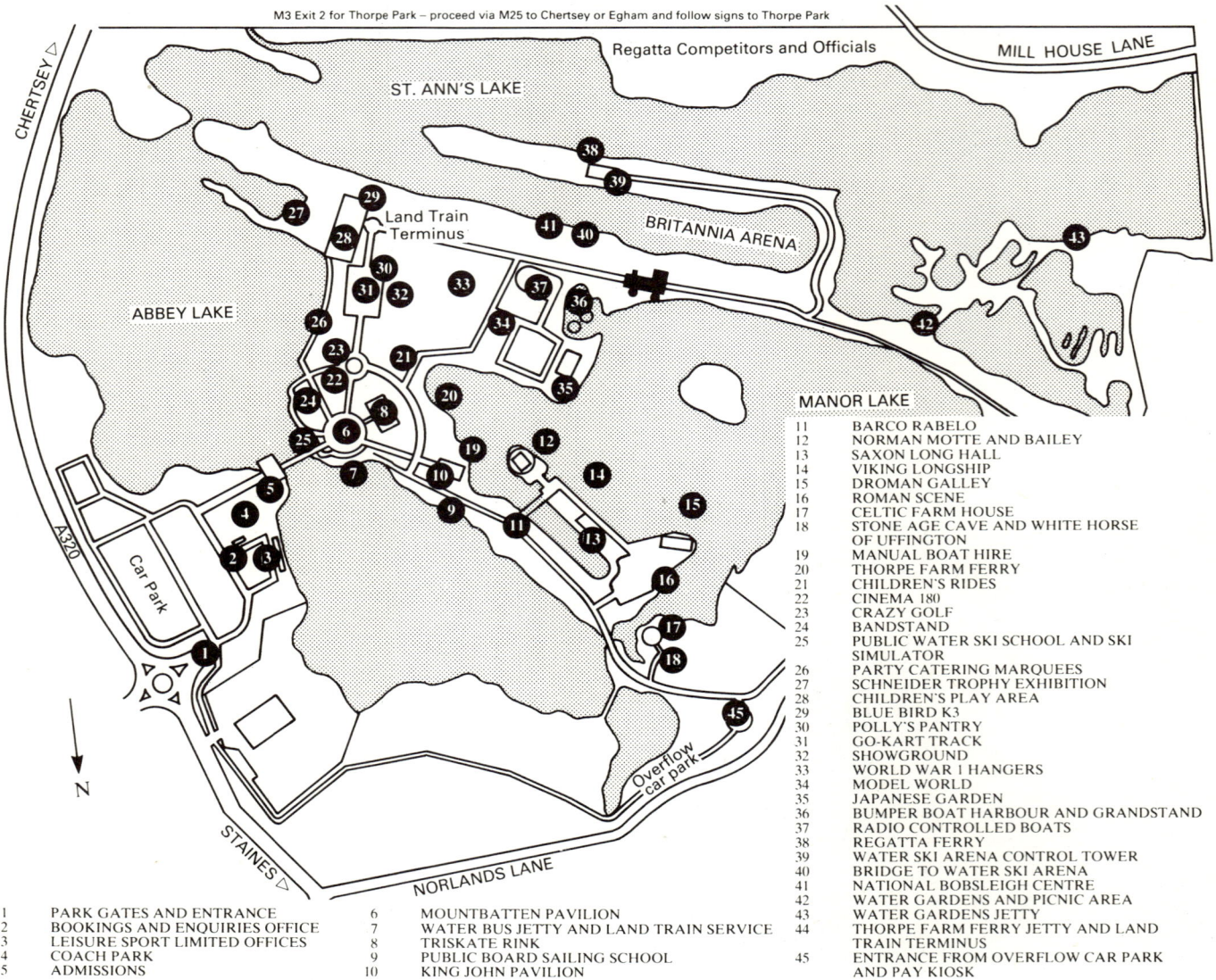

M3 Exit 2 for Thorpe Park – proceed via M25 to Chertsey or Egham and follow signs to Thorpe Park

1	PARK GATES AND ENTRANCE
2	BOOKINGS AND ENQUIRIES OFFICE
3	LEISURE SPORT LIMITED OFFICES
4	COACH PARK
5	ADMISSIONS
6	MOUNTBATTEN PAVILION
7	WATER BUS JETTY AND LAND TRAIN SERVICE
8	TRISKATE RINK
9	PUBLIC BOARD SAILING SCHOOL
10	KING JOHN PAVILION
11	BARCO RABELO
12	NORMAN MOTTE AND BAILEY
13	SAXON LONG HALL
14	VIKING LONGSHIP
15	DROMAN GALLEY
16	ROMAN SCENE
17	CELTIC FARM HOUSE
18	STONE AGE CAVE AND WHITE HORSE OF UFFINGTON
19	MANUAL BOAT HIRE
20	THORPE FARM FERRY
21	CHILDREN'S RIDES
22	CINEMA 180
23	CRAZY GOLF
24	BANDSTAND
25	PUBLIC WATER SKI SCHOOL AND SKI SIMULATOR
26	PARTY CATERING MARQUEES
27	SCHNEIDER TROPHY EXHIBITION
28	CHILDREN'S PLAY AREA
29	BLUE BIRD K3
30	POLLY'S PANTRY
31	GO-KART TRACK
32	SHOWGROUND
33	WORLD WAR 1 HANGERS
34	MODEL WORLD
35	JAPANESE GARDEN
36	BUMPER BOAT HARBOUR AND GRANDSTAND
37	RADIO CONTROLLED BOATS
38	REGATTA FERRY
39	WATER SKI ARENA CONTROL TOWER
40	BRIDGE TO WATER SKI ARENA
41	NATIONAL BOBSLEIGH CENTRE
42	WATER GARDENS AND PICNIC AREA
43	WATER GARDENS JETTY
44	THORPE FARM FERRY JETTY AND LAND TRAIN TERMINUS
45	ENTRANCE FROM OVERFLOW CAR PARK AND PAY KIOSK

Fig 1.27 ● Thorpe Park

Garden Paths

Savill Garden

Here is the official map for the Savill Garden at Windsor Great Park.

1. Roughly how far is it (in feet) along the Obelisk Ride? (The answer is one of these: 400 feet, 800 feet, 1400 feet, 500 feet, 2000 feet.)

2. Look for the arrow pointing north. Would you go roughly north (sun behind you), south (into the mid-day sun), east or west if you went from the Obelisk Pond to the nearest toilets?

3. Which compass direction would you follow to go from the Obelisk Pond to the Jubilee Garden?

4. Give directions for going from the Restaurant to see the Dry Garden (keeping to the paths all the way).

5. Roughly how far is it from the Moisture Loving Plants to the Temperate House?

Fig 1.28 ● Savill Garden

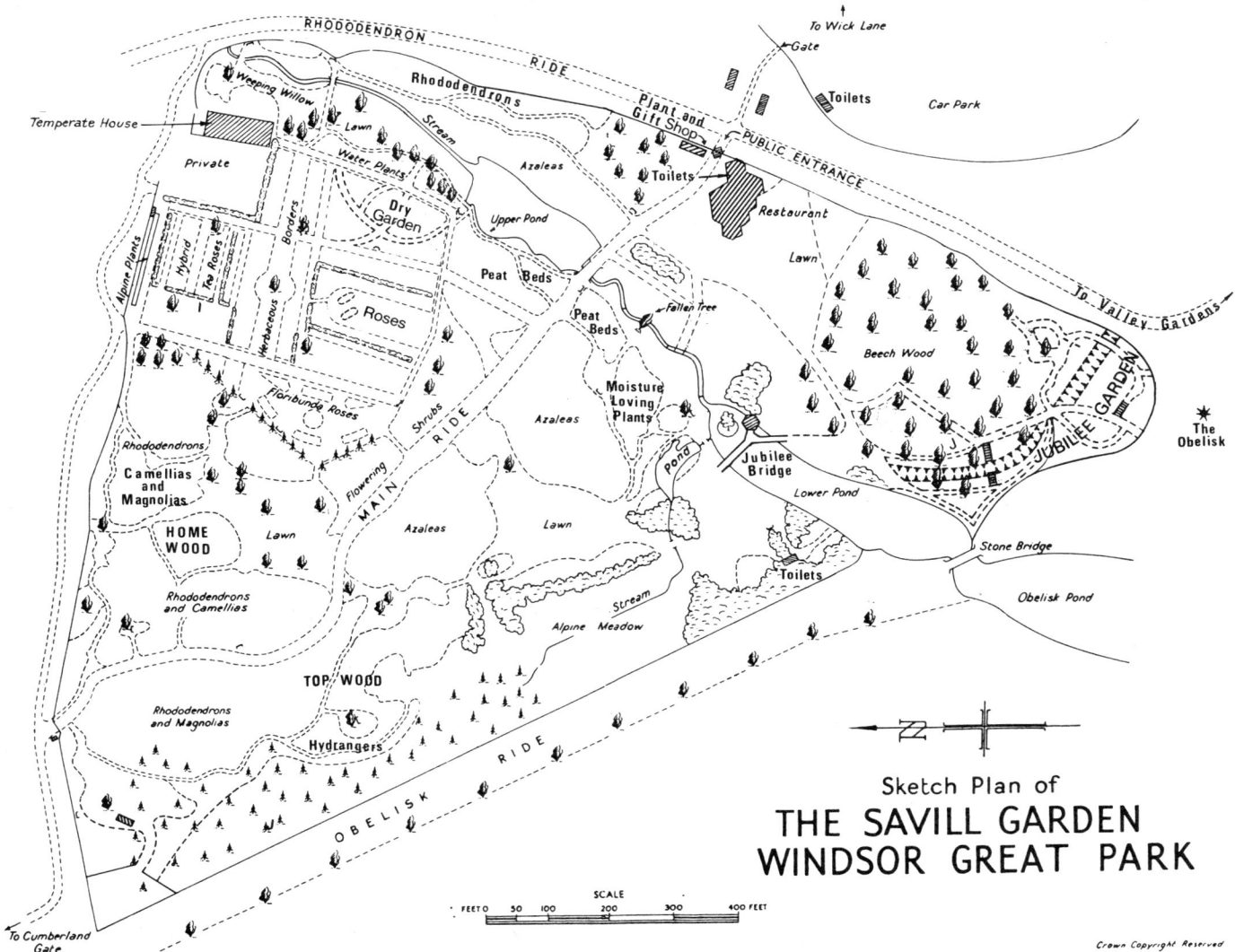

Sketch Plan of
THE SAVILL GARDEN
WINDSOR GREAT PARK

Distance Charts

Have you ever tried to work out how far it is from one place to another on a map? This is easy enough on a straight line, but not many roads are straight! How do you know how much to allow for the length of each bend? And does the measurement you obtain fit neatly onto the scale shown on the map?

A distance chart does not take much space, but contains a great deal of information. It shows the real distances between towns and allows you to choose which ones you compare. There is a small chart shown on this page but even this short one contains forty-five items of information. Most charts in books show more towns than this and contain many hundreds of items of information.

Distance in km

Fig 1.29 ● Distance Chart

Distance charts can be helpful. They tell you how far you have to go, and therefore roughly how long it will take. They are used for mileage claims (you can work out the amount to charge for the cost of a journey). They can help also with route planning, deciding that it is shorter to go by one route rather than another.

Use this chart to answer these questions:

1. How far is it from London to York?
2. Which is further, from Hull to Norwich, or from Hull to Liverpool?
3. Which is further, from Aberdeen to Manchester, or from Manchester to London?
4. Is Liverpool nearer to Bournemouth or to Harwich?
5. Is Shrewsbury further from Hull or from Bournemouth?
6. Sadie lives in Shrewsbury, and has a cousin in Hull and a good friend in Norwich. She wants to see one of them this weekend, but does not want to make a very long journey. Which one should she visit?
7. Melissa has been to an important meeting for her company, travelling from Bournemouth to Manchester and back. She can claim 25p a mile (£1 for each four miles travelled). How much does she claim?
8. Claudia is a saleswoman from Italy driving a Fiat with distances shown in km. She usually travels at an average speed of 80 km/h on long distance journeys. How long must she allow to travel from York to Aberdeen, allowing an extra two hours for breaks?

Scale

The maps and illustrations on the next page show the same place on different scales. Large-scale maps give details that are missed on small-scale maps. However, small-scale maps can give information that cannot fit on large-scale maps.

1. Name some useful details that you can find on the large-scale map that are not found on the other maps.
2. Why are the roads shown in Fig 1.32 not all shown in Fig 1.33?
3. Why is Slimbridge not shown as an ordinary place name in Figs 1.32 and 1.33?
4. Which map would be most useful for someone who lived (a) in Bristol (b) in Preston, and who wanted to visit Slimbridge?

Scale

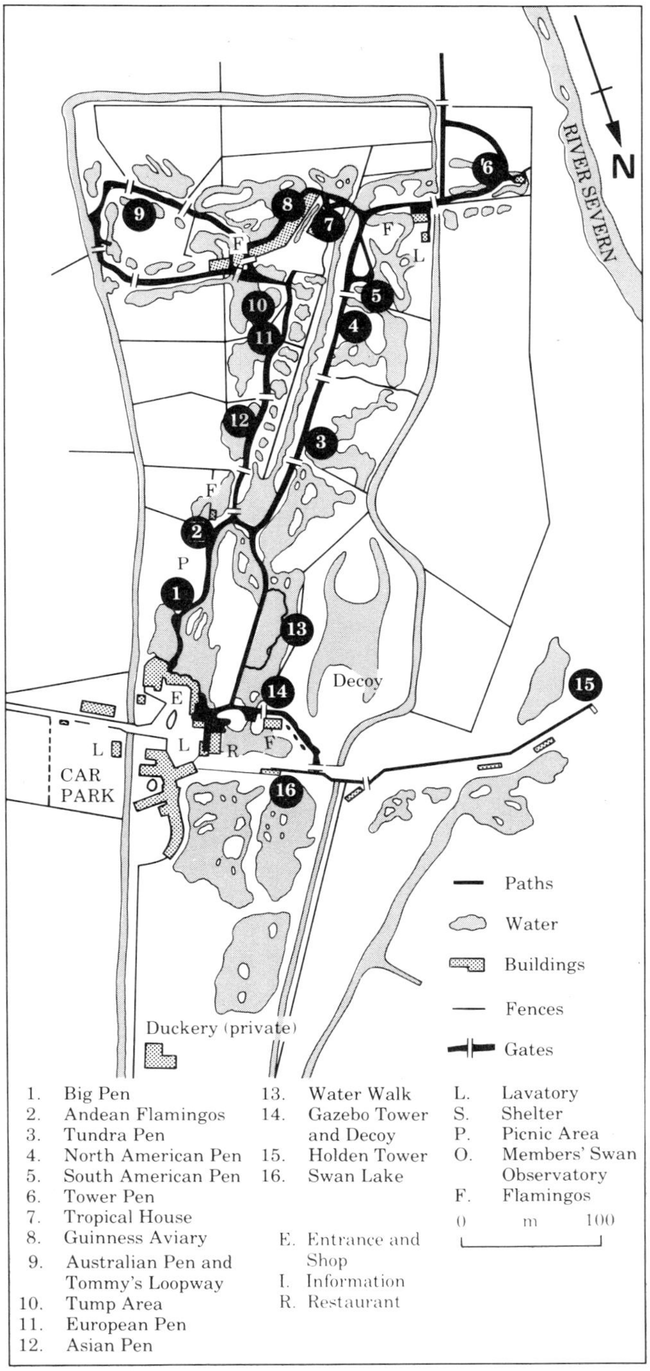
RIVER SEVERN
N
Decoy
CAR PARK
Duckery (private)
Paths
Water
Buildings
Fences
Gates
1. Big Pen
2. Andean Flamingos
3. Tundra Pen
4. North American Pen
5. South American Pen
6. Tower Pen
7. Tropical House
8. Guinness Aviary
9. Australian Pen and Tommy's Loopway
10. Tump Area
11. European Pen
12. Asian Pen
13. Water Walk
14. Gazebo Tower and Decoy
15. Holden Tower
16. Swan Lake
E. Entrance and Shop
I. Information
R. Restaurant
L. Lavatory
S. Shelter
P. Picnic Area
O. Members' Swan Observatory
F. Flamingos
0 m 100

Fig 1.30 ● Large scale map of Wildfowl Trust

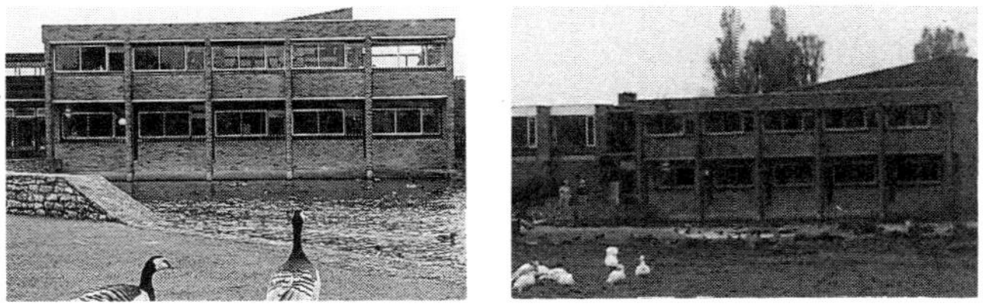

Fig 1.31 ● Large scale (left) and small scale (right) photographs of entrance

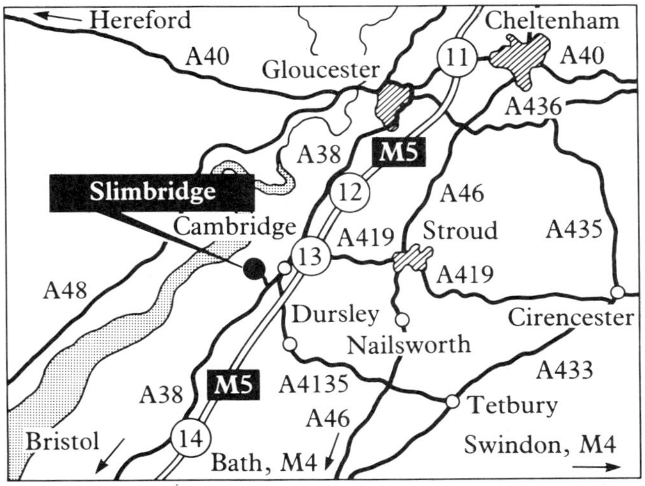
Hereford
A40
Gloucester
Cheltenham
A40
A436
A38
M5
Slimbridge
Cambridge
A46
Stroud
A435
A419
A419
A48
Dursley
Nailsworth
Cirencester
A433
A38
M5
A4135
Tetbury
A46
Bristol
Bath, M4
Swindon, M4

Fig 1.32 ● Routes to Slimbridge from 30 miles away

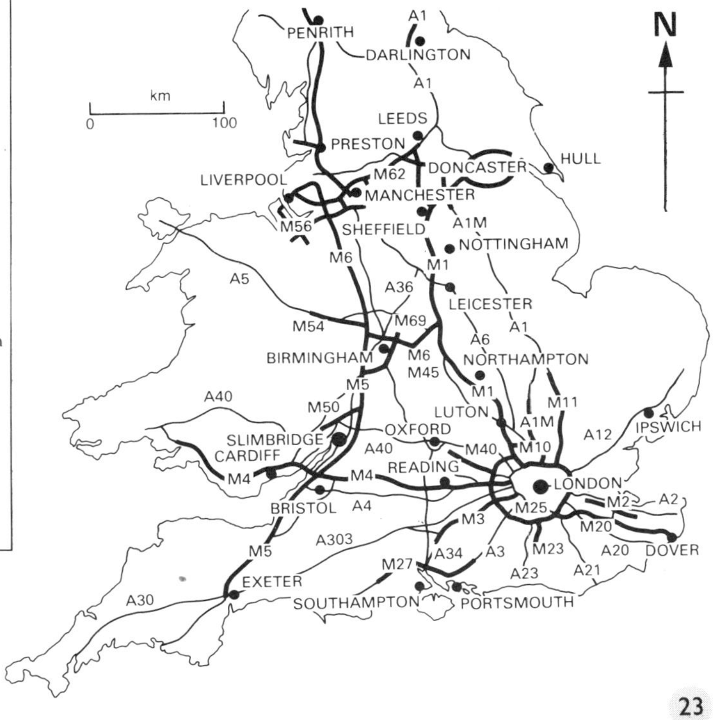
N
km
0 100
PENRITH
A1
DARLINGTON
LEEDS
PRESTON
HULL
LIVERPOOL
M62
DONCASTER
MANCHESTER
M56
SHEFFIELD
A1M
NOTTINGHAM
M6
M1
A5
A36
LEICESTER
M54
M69
A1
BIRMINGHAM
A6
M6
NORTHAMPTON
M45
M5
M1
A40
M50
LUTON
M11
A1M
SLIMBRIDGE
OXFORD
IPSWICH
CARDIFF
A40
M40
M10
A12
M4
M4
READING
LONDON
BRISTOL
A4
M25
M2
A2
M3
M20
M5
A303
A34
A3
M23
A20
DOVER
EXETER
M27
A23
A21
A30
SOUTHAMPTON
PORTSMOUTH

Fig 1.33 ● Routes to Slimbridge from up to 200 miles away

Compass Directions

The four points of the compass are parts of our daily language, because they are such a useful way of describing where a place is. News reports speak of 'the North of England', 'in the South', 'the East Coast', or 'the West Country'. The words north, south, east, west appear in weather forecasts, countless street and place names and a large number of trade names as well.

Midway between each of these four points lie north-east, south-east, south-west and north-west. These words are also used commonly.

A rough idea of compass direction can be found by remembering these facts:

North can be found by looking on a starlit night for the Pole Star.

East is the direction to look to see the sun rise.

South is the direction to look to see the sun at midday.

West is the direction to look to see the sun set.

(The sun is only exactly in these positions on one day in spring and one day in autumn.)

On a compass, it is important to check that the needle can move freely. Compasses used in cars are mounted in dome shaped bowls to allow this freedom on hills and uneven road surfaces.

Navigators use more exact compass positions, but they do not concern us here.

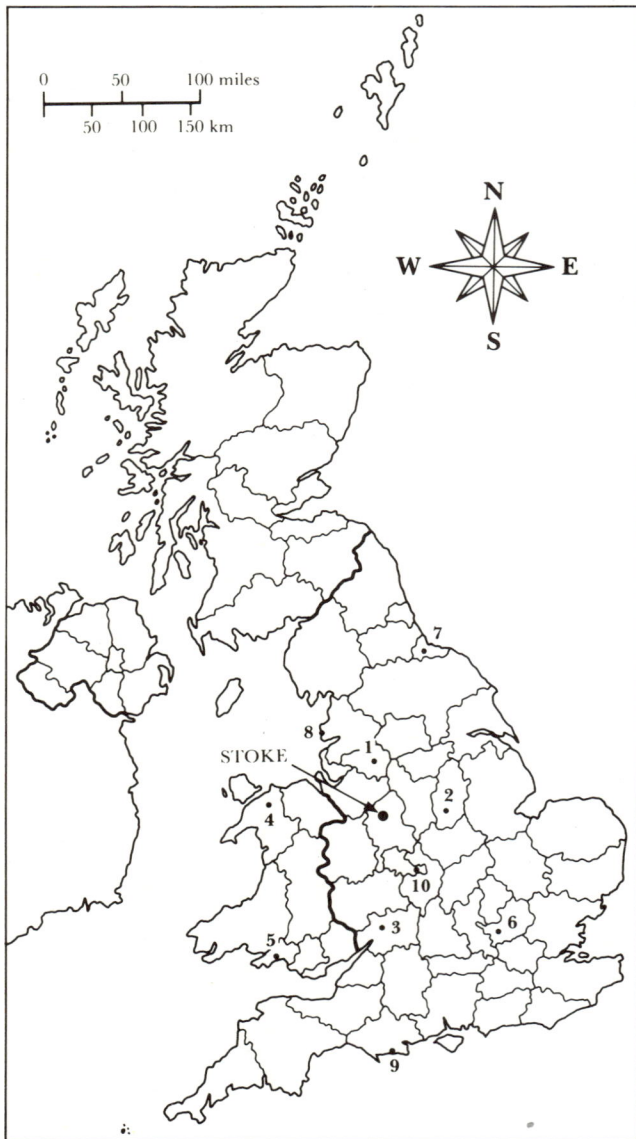

● Compass

Fig 1.34 ● Key to places listed

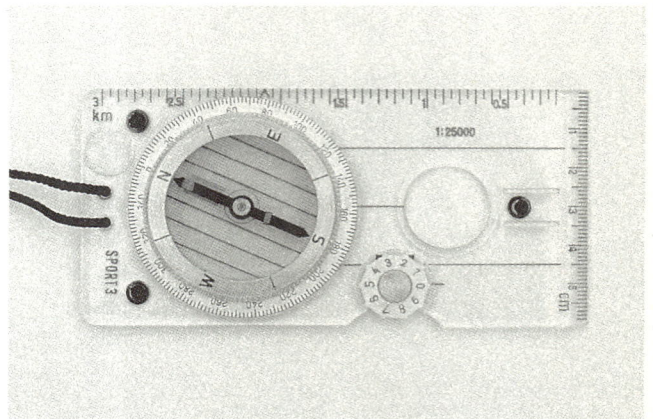

Which Way to These Places?

Using the map of Great Britain, say in which compass direction each of these places lies if you are in Stoke-on-Trent (shown with the big black circle):

1. Manchester
2. Nottingham
3. Gloucester
4. Snowdon
5. Swansea
6. Hemel Hempstead
7. Middlesbrough
8. Blackpool
9. Bournemouth
10. Coventry

Norfolk Lanes

Here is a map of the lanes near Fakenham in Norfolk. Follow these instructions from Barmer Church to the church at North Creake.

- Go along the main road south-east for about 100 metres.
- Turn north-east onto a lane through Barmer.
- Keep on this lane for just over 2 km and then turn north.
- Keep on this path for nearly 3 km and then turn north-east.
- Keep on this path for about 2 km and turn east on to the road.
- After less than half a kilometre, turn right onto the main road and you will come to North Creake Church.

1. What named points did you pass on the way?

2. What would have been the problem with finding your way on this route if you did not use compass bearings? Would you expect to find many signs in this district?

3. Why would it not be good enough to rely on using a watch as a compass on this journey? What sort of weather could present some dangers?

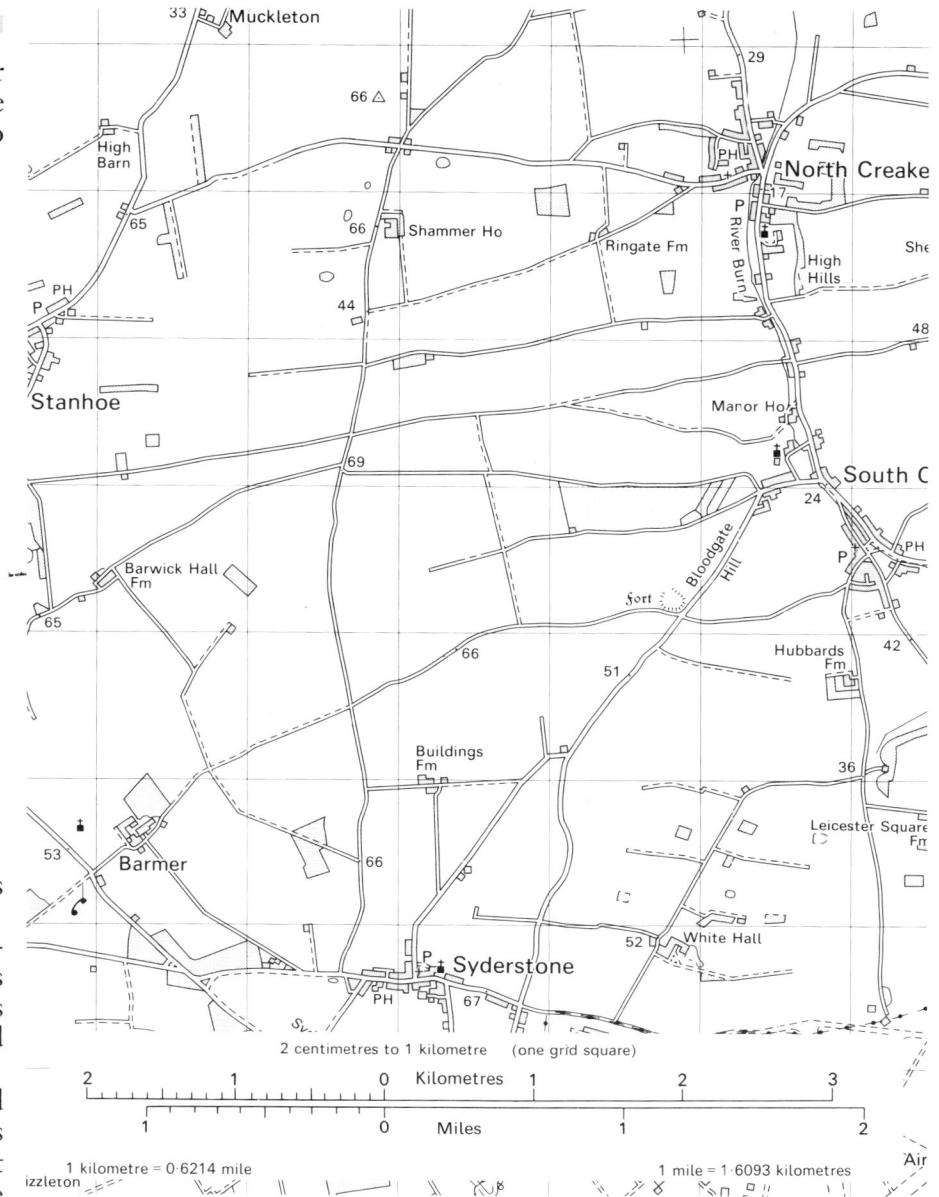

Fig 1.35 • Ordnance Survey map of country near Fakenham, Norfolk. (*Reproduced from the 1987 Ordnance Survey 1:50 000 Landranger map with the permission of the Controller of H.M. Stationery Office Crown copyright reserved*)

Taking Bearings

Stand in an open space, such as a field, and take bearings. What can you see at each of the eight points of the compass named in figure 1.36? Are these objects near, middle distance, or far away?

Compass Directions

Prepare instructions for someone else in your group to make a short journey in your own district following compass directions. If you follow such a route, do you have any difficulties or problems?

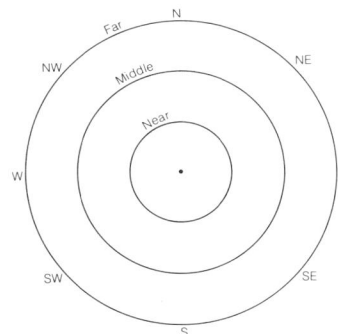

Fig 1.36 • Bearings diagram

Timetables

24-Hour Clock

Most timetables are printed using the 24-hour clock. This stops confusion about whether the time given is in the morning or the afternoon. These times are always shown with four digits. Seven o'clock in the morning is written 0700 and quarter past seven in the morning is written 0715. After midday, the hours are counted on to the twelve hours that have already passed. One o'clock in the afternoon becomes 1300, seven o'clock in the evening becomes 1900 (12 + 7), and quarter past seven in the evening becomes 1915.

Fig 1.37 ● A 24 hour clock

Flights to and from Cardiff

The timetable here shows some flights between four airports and Cardiff. Some flights are not made every day and some special times apply on certain dates.

1. Which of these airports is not in the United Kingdom?
2. Where does flight DA806 go to? What time does it take off and when does it arrive?
3. How many days a week does flight DA125 depart for Belfast?
4. If you wanted to come back from Jersey to reach Cardiff before six in the evening, which flight number would you take?
5. If you wanted to come back from Amsterdam Schiphol to reach Cardiff by 12.00, what time would you leave Amsterdam?
6. Can you fly from Guernsey to Cardiff on Tuesdays?
7. How long does the 0700 flight from Cardiff to Amsterdam appear to take?
8. How long does the 2000 flight from Amsterdam to Cardiff appear to take? Can you explain why there seems to be a difference between your answers for questions 7 and 8?

Days of Service		Dep.	Arr.	Flight	Cl	a/c

From Cardiff-Wales
To AMSTERDAM Schiphol International

Days of Service		Dep.	Arr.	Flight	Cl	a/c
	1	0700	1005	DA800	YM	HS7
A	3 5	0700	1005	DA800	YML	HS7
B	3 5	0700	1005	DA800	YML	HS7
C	1 3 5	1530	1840	DA808	YML	148
B	1 3 5	1530	1840	DA808	YML	148

A Until 20 Dec   B From 27 Dec
C Until 23 Dec

To BELFAST International

		Dep.	Arr.	Flight	Cl	a/c
A	12345	1120	1335	DA125	YML	HS7
B	12345	1120	1335	DA125	YML	HS7

A Until 24 Dec   B From 27 Dec

To GUERNSEY

		Dep.	Arr.	Flight	Cl	a/c
A	1 3 5	1215	1300	DA263	YM	146
B	1 3 5	1215	1300	DA263	YM	146

A Until 23 Dec   B From 27 Dec

To JERSEY

		Dep.	Arr.	Flight	Cl	a/c
A	7	1200	1300	DA355	YM	HS7
B	7	1200	1330	DA257	YM	HS7
C	1 3 5	1215	1345	DA263	YM	148
D	1 3 5	1215	1345	DA263	YM	148

A Until 8 Dec   B From 15 Dec
C Until 23 Dec   D From 27 Dec

Days of Service		Dep.	Arr.	Flight	Cl	a/c

To Cardiff-Wales
From AMSTERDAM Schiphol International

		Dep.	Arr.	Flight	Cl	a/c
A	1 3 5	1035	1145	DA801	YML	146
B	1 3 5	1035	1145	DA801	YML	146
A	1 3	2000	2115	DA807	YML	HS7
C	1 3	2000	2115	DA807	YML	HS7
	5	2000	2115	DA807	YM	HS7

A Until 23 Dec   B From 27 Dec
C From 30 Dec

From BELFAST International

		Dep.	Arr.	Flight	Cl	a/c
A	12345	0925	1055	DA125	YML	HS7
B	12345	0925	1055	DA125	YML	HS7

A Until 24 Dec   B From 27 Dec

From GUERNSEY

		Dep.	Arr.	Flight	Cl	a/c
A	1 3 5	1320	1455	DA263	YM	146
B	1 3 5	1320	1455	DA263	YM	146

A Until 23 Dec   B From 27 Dec

From JERSEY

		Dep.	Arr.	Flight	Cl	a/c
A	7	1400	1545	DA258	YM	HS7
B	1 3 5	1410	1455	DA263	YM	146
C	1 3 5	1410	1455	DA263	YM	146
D	7	1630	1730	DA366	YM	HS7

A From 15 Dec   B Until 23 Dec
C From 27 Dec   D Until 8 Dec

Fig 1.38 ● Flights between Cardiff and four airports

By Rail to France and Italy

The railway timetable shows times from London to Italy. Train timetables show some of the stations you can reach, but you may have to change trains to do so. A solid line down the page shows that you can stay on the same train, and the times for that train are printed in ordinary type. If you want to reach a station where the times are in *italics*, you must change at the last station with times shown in ordinary type. Rows of small dots show the trains do not serve these stations.

> 1. If you want to go from London Charing Cross to reach Paris Nord by 1700 hrs, what time must you set off?
> 2. If you want to go from London Waterloo to reach Chambery by midnight, what time must you set off?
> 3. If you want to go by sea rather than by hovercraft, which London station do you need?
> 4. The train that reaches Paris St-Lazare at 1807 is a few minutes behind the one that reaches Paris Nord at 1740. What time did the passengers on each train leave London? Why did one take three hours more than the other?
> 5. Where would you change train to reach Rimini by 0939 next morning?
> 6. If you want to reach Firenze SMN without changing train, which route number would you choose?

Which Way to Travel?

Here are extracts from timetables for coaches, trains and planes between Plymouth and Aberdeen. Make your own comments about the advantages and disadvantages of each route. Remember to think about the comfort and convenience of the passengers, the choice of times, the effects of possible delays, the length of the journey, and the pleasures of travelling including what passengers might see on the way.

Air:
One flight daily,

Depart Plymouth	1720
Arrive Aberdeen	2020

Coach:

Depart Plymouth:	0650	1540	2359
Change: London		2005	0500
Change: Glasgow	1630		
Arrive Aberdeen:	2110	0835	2230

Rail:

Depart Plymouth:	0540	0700	0833	0945	1718
Change Bristol:	0815				
Change York:			1458		2329
Change Newcastle:		1404			
Change Edinburgh:	1453				
Change Dundee:		1806			
Arrive Aberdeen:	1730	1946	2030	2200	0700

Fig 1.39 ● Timetables between Plymouth and Aberdeen

SEE NOTE		A	B	C	D	E
London Victoria	d.	...	...	...	...	08 04
London Charing Cross	d.	09 55*	10 55	....	...	....
London Waterloo (East)	d.	09 57*	10 57	...	...	....
SEA CROSSING		⛴	⛴			⛴
Newhaven	d.	....	....	....	....	10 00*
Folkestone Harbour	d.	...	...	...	...	...
Dover Western Docks	d.	....	....	....	....	....
Dover Eastern Docks	d.	...	...	....	...	...
Dover Hoverport	d.	12 10*	13 10	....	....	....
TRAIN NUMBER		2022	2028			308
Oostende	d.	...	...	...	...	...
Calais Maritime	d.	....	....	....	....	...
Boulogne Maritime	d.	...	...	...	...	...
Boulogne Hoverport	d.	14 10	15 10	...	...	...
Dieppe Maritime	d.			...	...	15 48
Bruxelles Midi	a.	...	...	...	...	...
Paris St-Lazare	a.			...	...	18 07 -
Paris Nord	a.	16 32	17 40			
TRAIN NUMBER				213	221	223
Bruxelles Midi	d.	....	...	...	...	...
Paris Est	d.	....	...	...	...	...
Paris Lyon	d.			18 47	18 50	19 32
Chambéry	a.	....	....	23 51		
Basel SBB	a.	...	...			
TRAIN NUMBER				23 59		
Chambéry	d.	....	....	23 59		
Basel SBB	d.	...	...			
Torino P.N.	a.	....	....			
Torino P.S.	a.	....	....			
Milano Lambrate	a.	....	....		03 57	
Milano Centrale	a.	....	....			
TRAIN NUMBER						
Milano Centrale	d.	....	....			
Verona P.N.	a.	...	...		05 29	
Venezia Mestre	a.	...	...		06 57	
Venezia S.L.	a.	...	...		07 20	
Trieste Centrale	a.	...	...		09 15	
Bologna Centrale	a.	...	...		...	07 16
Rimini	a.	...	...		...	*09 39*
Ancona	a.	...	...		...	*11 20*
Brindisi	a.	...	...		...	...
Genova P.P.	a.	...	...		...	...
Pisa Centrale	a.	...	...	06 59	...	...
Firenze S.M.N.	a.	...	...	*08 25*	...	08 42
Roma Tiburtina	a.	...	...		...	...
Roma Termini	a.	...	...	10 05	...	...
TRAIN NUMBER				2817		
Roma Termini	d.	....	...	12 15		
Napoli Centrale	a.	...	...	14 45		...

Fig 1.40 ● Rail times from London to Italy

Fig 1.41 ● Swindon coach station

Time-Distance

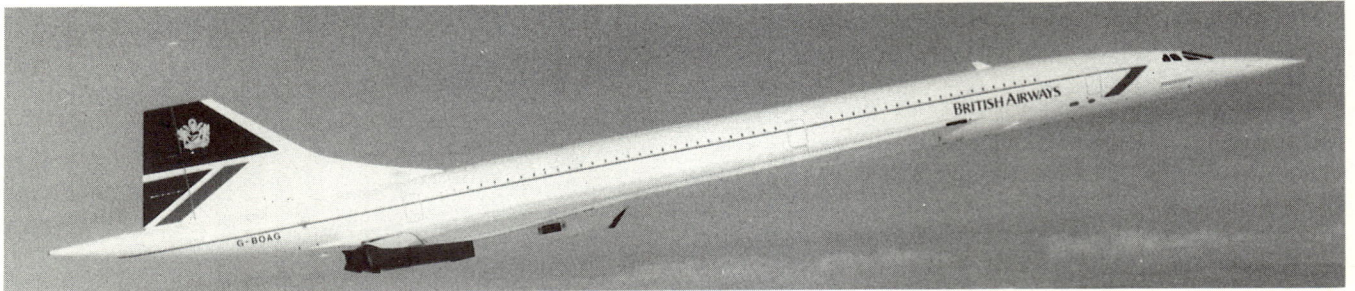

Fig 1.41 ● Concorde

It is not always important to know how far away a place is. It may matter much more to know how long it takes to get there. For this reason, many maps about public transport give the distance in time and not in miles.

Compare these two maps of the world. On the one on the left, the scale is in time-distance. The one on the right shows the distance in miles.

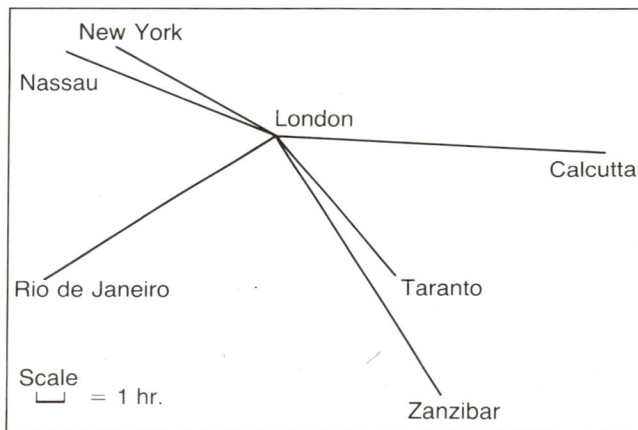

Fig 1.42 ● Time-distance map

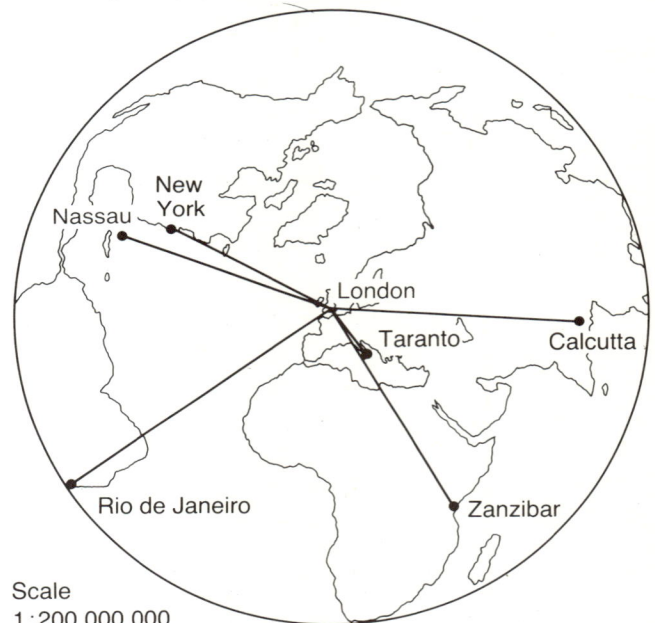

Fig 1.43 ● Distance in miles

1. On map A, which is quicker to reach from London, New York or Taranto in southern Italy?
2. On map B, how much shorter is the journey from Taranto to London than the journey from New York to London?
3. Why does it take so much longer to reach Taranto from London (via Naples) than to reach New York from London?
4. On map A, which is quicker to reach from London, Zanzibar (part of Tanzania) or Nassau in the Bahamas?
5. What is the difference between the distance in miles from London to Tanzania and the distance from London to the Bahamas?

6. Why does it take so much longer to reach Zanzibar than to reach the Bahamas?
7. On map A, which is quicker to reach from London, Calcutta or Rio de Janeiro?
8. On map B, which is nearer to London, Calcutta or Rio de Janeiro?
9. In conclusion, what things make a difference in time-distance that are not linked with the difference in miles?

Time Zones

Timetables for long distance journeys usually show the time of arrival in local time. A plane that leaves London at midnight and reaches Los Angeles at 5 a.m. has not really taken only five hours. Los Angeles time is seven hours behind British time, so the flight has really taken twelve hours.

This map shows the different time zones across the world. Places at the same longitude may have different times because of political borders.

1. If the time is noon in London, what time is it in:

- Kiev?
- Vienna?
- Helsinki?
- Istanbul?
- Accra?
- Auckland?
- Buenos Aires?
- Hong Kong?
- Vancouver?

2. The box (right) shows an extract from the British Telecom Dialling Codes booklet. Using the Telecom information, which two countries shown in the table have the same time as Britain? Can you find out where these countries are?

3. What would be the time in the Marshall Islands if you phoned them at 1500 hrs British time?

4. What would be the time in Kuala Lumpur if you phoned at 1500 hrs British time? Where is Malaysia?

Country	Time diff. in hrs.
Malaysia	+ 7½/8
Kota Kinabalu	
Kuala Lumpur	
Maldives	+ 5
Mali	0
Malta	+ 1
Marshall Islands	+ 12
Martinique	− 4
Mauritania	0
Mauritius	+ 4
Mexico	− 6
Acapulco	
Micronesia	+ 9/11
Midway Island	− 11
Monaco	+ 1
Montserrat	− 4

Fig 1.44 ● World times related to Greenwich Mean Time (GMT). The map shows approximate time zones only. The actual lines follow the borders of countries and may vary with 'Summer Time' clock changes

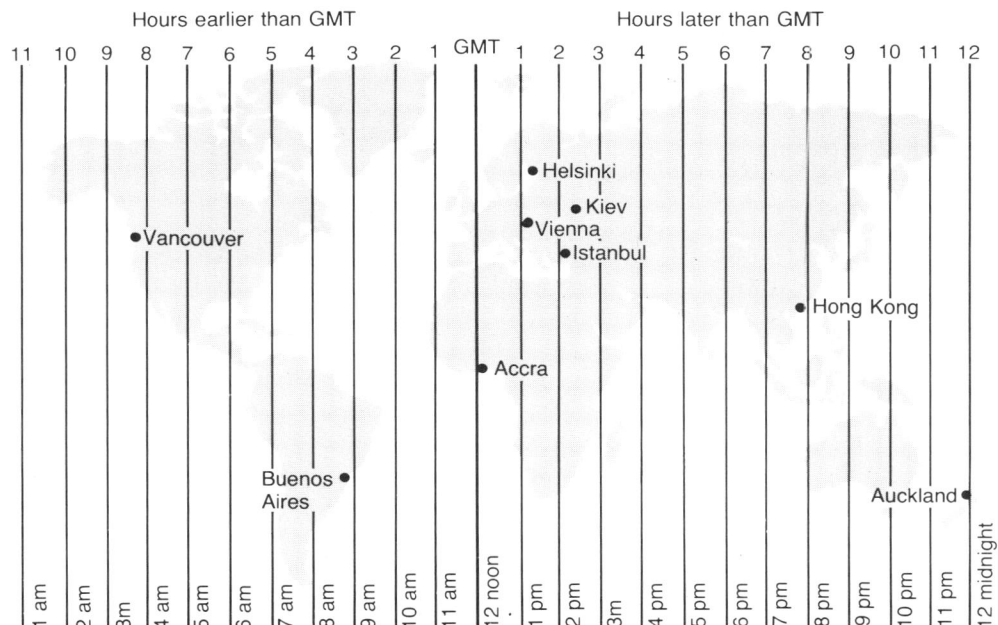

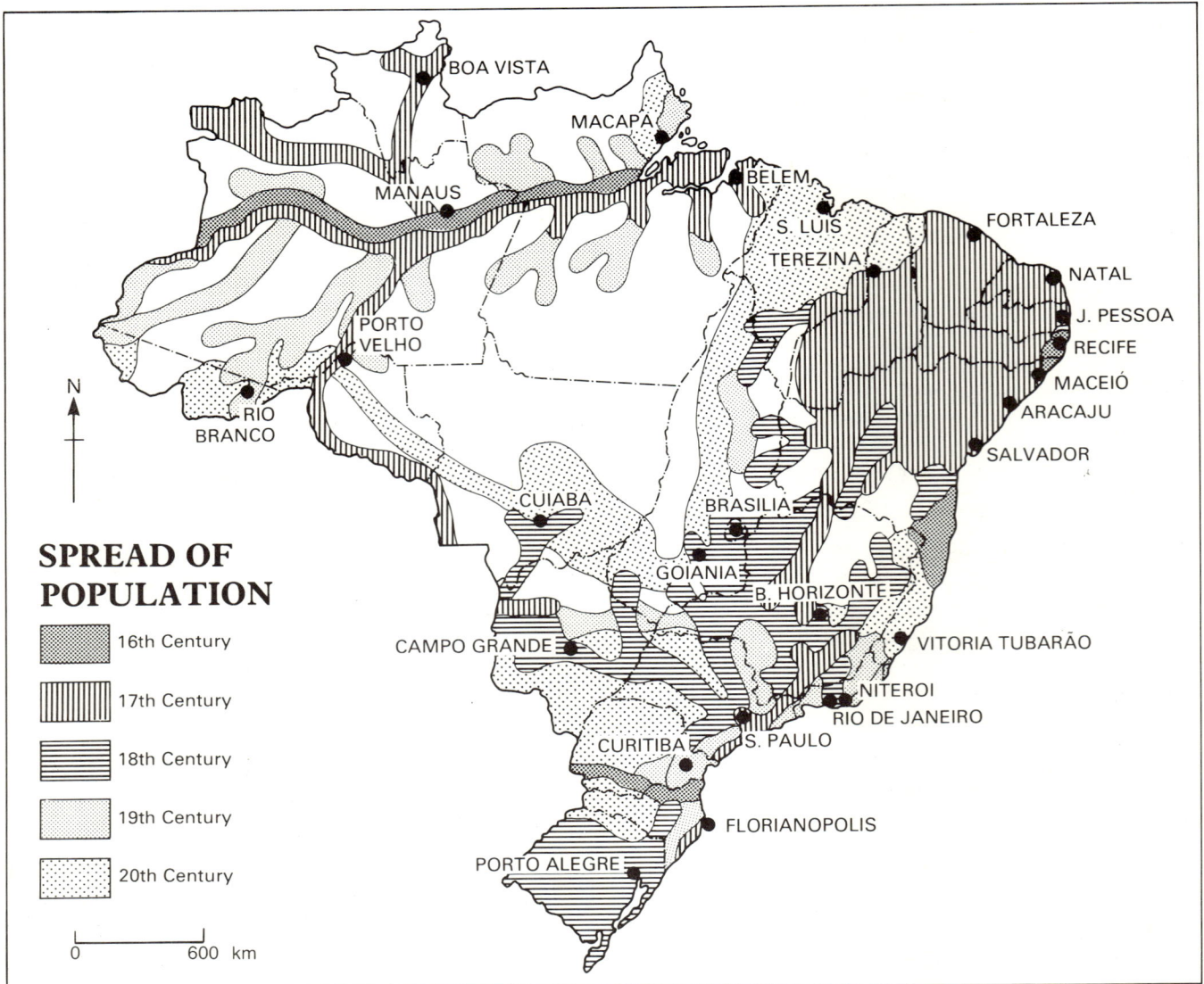

SPREAD OF POPULATION

	16th Century
	17th Century
	18th Century
	19th Century
	20th Century

0   600 km

Fig 1.45 ● Population settlement of Brazil

Brazil

The map shows the way people have moved into the vast country of Brazil. Although there are now more than 120 million people, many parts of the country are not heavily populated. Brazil's government used to encourage people to have large families to help fill these areas. The policy now is to encourage limiting the size of families.

1. The map makes it look as if Brazil were an island. Look at an atlas and check which borders are coastline, and which are borders with other countries?
2. Where did the first people settle in Brazil in the sixteenth century and in the seventeenth century?
3. Brasilia (in the interior) has been developed as the new capital city of the country, replacing Rio de Janeiro (near the coast). When was the area around Brasilia first settled?

German Rescue Services

These maps are taken from a booklet sold in West Germany. Similar information is often given on maps all over the world to help you find services that you need.

The maps show the services provided rather than all the towns, roads and hills. In order to be able to use the maps, you need to be able to find where you are. This can be done by following the bends of a river, or by knowing that you are a certain distance between two big towns that are marked. Coastlines and national frontiers can help to show where you are.

1. What is the telephone number for emergency services in Bremen?
2. What sort of emergency service is available at
   a) Bielefeld   b) Osnabruck?
3. What broadcasting area is Köln in?
4. What is the frequency of the short wave transmitter nearest to Köln?
5. If you were in a town north-east of Koblenz, which three radio areas might you be able to hear?

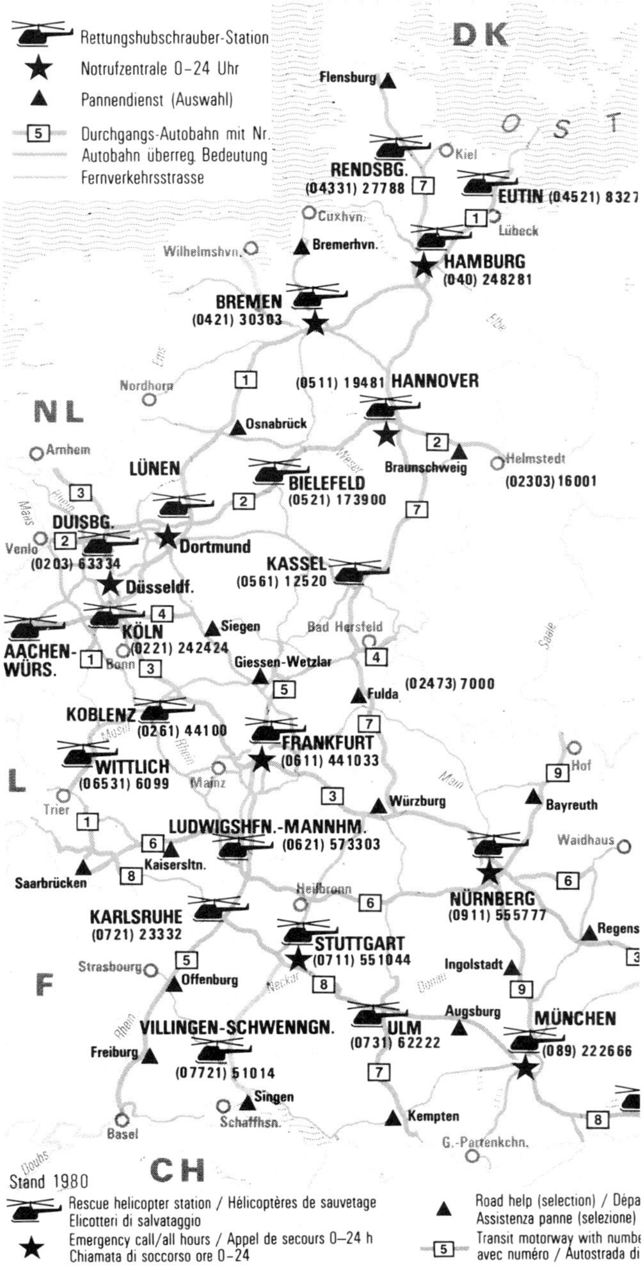

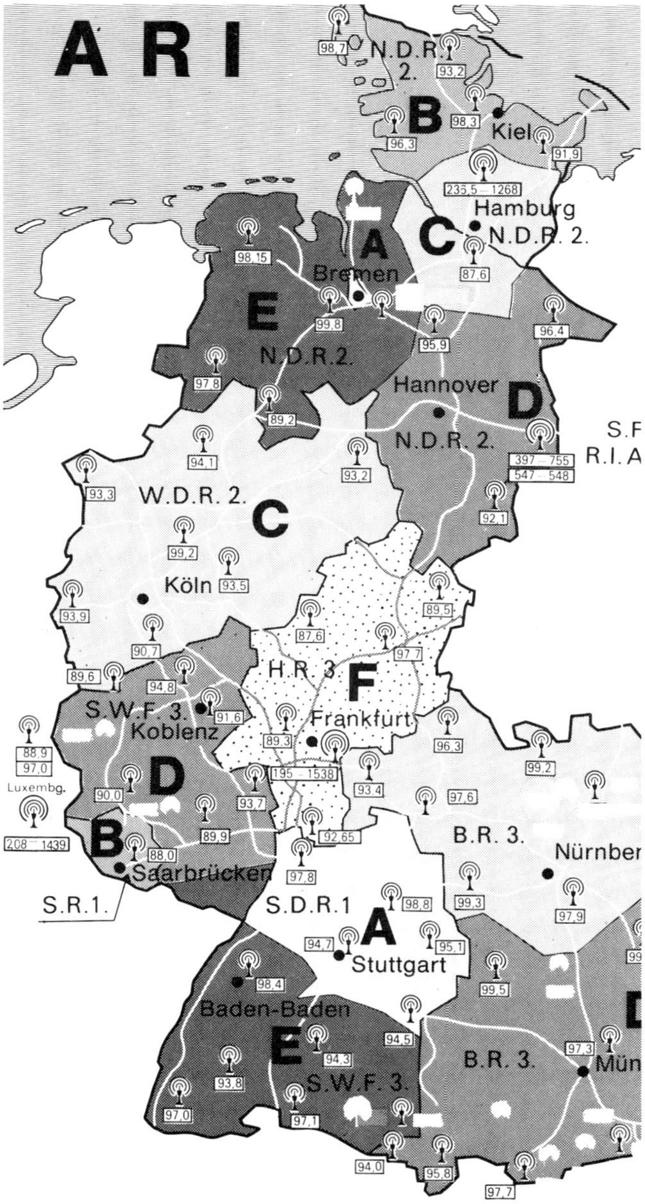

Fig 1.46 ● (left) Part of German breakdown services map

Fig 1.47 ● (above) Part of German broadcasting map

Quiz

Spot the Errors

What is not correct in the following sentences?

1. A street plan is better than a photograph because it is to scale, shows every street, and helps you to recognise particular buildings.
2. On all street plans that mark co-ordinate grids, the letters go from left to right along the bottom, and the numbers go from zero up the side.
3. All Ordnance Survey maps are on the scale 1:50 000 (2 cm: 1 km).
4. The M3 lies to the north of Thorpe Park.
5. The AA print the map of Switzerland on a separate sheet because it is the only country in Europe without a coastline.
6. Ordnance Survey Landranger maps show details of hills, rivers, roads and climate.
7. If you want to know how far it is from one town to another on a distance chart, you add up all the figures in the right-hand column.
8. International flights in one direction seem to take longer than in the other direction because the planes fly so much more slowly into the wind.

Clever Clogs

1. What is the important difference between a topological map and most other route maps and physical maps?
2. How many Ordnance Survey Landranger maps are there to cover the whole of the United Kingdom?
3. What can you find out from an Ordnance Survey map if the contour lines are close together?
4. What is the German word for motorway (page 19)?
5. How does the AA manage to print the huge area of northern Scandinavia on a sheet that is about the same size as the other European maps?
6. How far apart are the bolder grid lines printed on an Ordnance Survey Landranger map?
7. When might you need a map on a scale bigger than 1:50 000?
8. Why do we have time zones, rather than simply setting our clocks by the midday sun?

Picture Search

● How did this photograph show the advantages of maps over photographs?

● What sort of task was linked with this school?

● What was explained on the page with this photograph?

Pick the Word

1. Which of these do you link with Chepstow? Royal Navy / horse racing / coal mines / fruit farming.

2. Which of these is coloured blue on an Ordnance Survey Landranger map? bridle path / trunk road / motorway / town street.

3. The street plan of Newbury given on page 6 does not show a railway / motorway / river / canal.

4. The best map to show details of hills and rivers is a town plan / physical map / main routes map / throughway map.

5. A topological map is particularly useful if you are travelling by car / bicycle / lorry / public transport.

6. The symbol ⊢——⊼——⊼ on an Ordnance Survey Landranger map shows a quarry / heath / demolition area / electricity transmission line.

7. At three in the afternoon, the sun will be roughly to the north-east / south-east / south-west / north-west.

8. The line that shows Greenwich Mean Time also passes through America / Africa / Australia / India.

Crossword

Make a tidy copy of this crossword grid and then solve the clues. Many of the main words are ones that appear on pages 2 to 31 of this book.

Clues across:

1. This is what we can measure if we know the scale of a map (8).
6. First three letters of the name 'Naomi' (3).
9. The way we often write 'north-west' (2).
10. The way we often write 'post office' (2).
11. The sort of map that is based on the survey organised from Southampton (8).
13. A little bit of something to try it out (6).
15. We need to know this if we are to work out the distance shown on a map (5).
16. The little word that tells us where a place is (2).
17. On one side, not on the main route (5).
19. The boundary between one country and another (8).

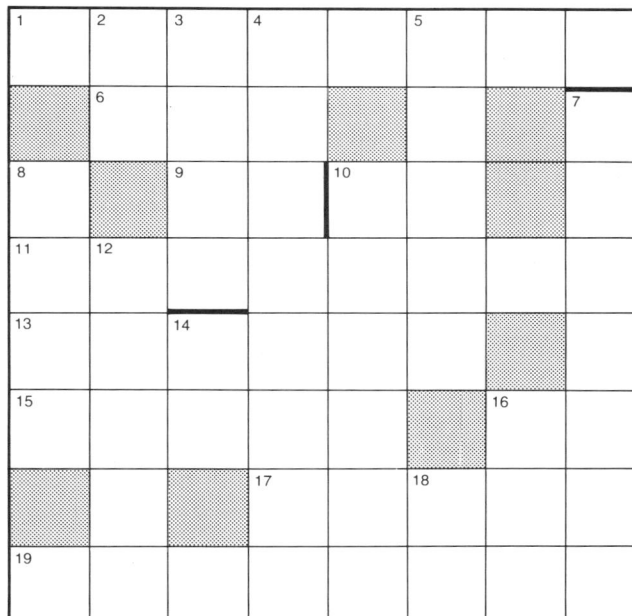

Clues down:

2. The little word that says a thing is not out of the area (2).
3. Many small dots along the coastline show that you might find this on the shore (4).
4. This sort of map helps you find your way around a town or city (4,4).
5. The opposite of 'everyone' (2,3).
7. An old city between Liverpool and North Wales, near the Wirral (7).
8. A town in Gwent at the end of the M50 (4).
10. The one that has the lightest colour is this (6).
12. A fast sort of bicycle (5).
14. First half of the word 'maps' (2).
16. Last three letters of the word 'lemonade' (3).
18. Vowels in the word 'limits' (2).

Studying the UK

Fig 2.1 ● Housing in Taunton. Is this similar to housing where you live?

Fig 2.2 ● Farmland in Kent. Is there landscape like this near where you live?

On a map of the world, the United Kingdom looks very small. Crowded into 230650 square km there are over 55 million people, and a great variety of landscapes, resources, and places of interest.

The United Kingdom is not the same thing as the British Isles. When we speak of the United Kingdom we are talking about those areas which are governed by the Government appointed by the Queen. That means England, Wales, Scotland, Northern Ireland and the many small islands extending from the Shetlands far in the north, the Hebrides in the north-west, to the Scilly Isles in the south-west and the Channel Isles near the French coast.

Eire (the Republic of Ireland) is part of the British Isles, but not part of the United Kingdom. Eire is shown on many maps of the United Kingdom, being the only foreign country bordering the UK.

The United Kingdom is one of the most crowded areas in the world and there is a lot to find out about it. It is always useful to know as much as possible about the country in which you live.

- You might hear of a job in another town. What would it be like there?
- You might make a friend while you are on holiday. What is it like where he or she lives?
- New neighbours might move into your street. Were things very different where they lived before?
- What is it like where your relations live?
- A team you support may be playing an away game. How far is it to go and watch them?
- The weather forecast tells you that snow is spreading from another district. How far away is that?
- You need to take the van a hundred miles away today. Will you be likely to be caught in fog? What forecast do you listen to?

This section of the book is organised to help you to find out and remember as many key facts as possible about the United Kingdom. You might want to make this a study project, building up a file or book of notes over a period of time. You might include postcards, leaflets, newspaper cuttings and other illustrations.

Starting from Home

Start with your own district.

- Which county is it in?
- What is the County Town?
- What sort of decisions does the Council make in its meetings there?
- What big towns are in your county?
- What other counties have borders that touch your county's borders?
- Is there a railway line in the district? If so, what are the next main stations in each direction? Where do the trains begin and end their journeys?
- What are the main roads running to and from your town?
- What are the next big towns in each direction along this road? What counties are they in?
- Is there a big river in your district? If so, where does it begin, and where does it join the sea?
- What towns are along this river? What counties does it flow through?
- What are the nearest big hills to where you live? How far is it from where you live to really high hills?
- How far is it from where you live to the sea? Which sea is it?

Making Short Descriptions

If you were far away from home and someone who had never been to your home district wanted to know what it was like, what would you say? You might like to describe:

houses Are they Tudor or Victorian, built of stone, wood or brick? New, or fifty years old? Large or small? Terrace rows or detached houses with big gardens?

factories Are they old and grimy, new and sparkling, large with tall chimneys, small and low, noisy or quiet?

landscape Do you have sweeping views of hills and woods, the sea or mountains? Is it flat, built up, or torn apart with mechanical diggers?

shops Are they small family shops serving local customers, or big department stores? Are they full of luxury goods which many people like to buy if they can afford them, or are the shops selling mainly basic things that people need to buy even when they have little to spend?

sports Are there big sports centres, tracks and grounds, theme parks, nature trails and picnic areas? Or are there only school playgrounds and back streets?

famous places Are there monuments, grand old buildings, landmarks and beauty spots that people come to visit?

farms Can you hear the lowing of cows and the bleat of sheep all day long, or have you silent fields? Is the land surrounded with thick hedges, stone walls or bare strands of wire?

These, of course, are only a few of the points to consider when describing an area. You will quickly think of more. By noting details of this sort, you will be able to pin down quickly the big differences between counties such as Cornwall and Cambridgeshire, Glamorgan and Cleveland.

Working Outwards

The next ten pages look at some of the best known places in the United Kingdom. To help you find your way around, the places have been grouped into five areas. These are:

- Scotland, Ireland, and North-east England
- London and the South-east
- West Midlands, North-west England, Mid and North Wales
- East Midlands, East Anglia, East Coast
- South and West Wales, West and South-west England

It is suggested that you should begin with the areas that you know. These will be the area in which you live, any other areas in which you have lived, areas you have visited and areas in which you have a special interest. Afterwards, try working through the points about areas you did not know about before. Think about where they are in relation to places that you do know.

When you have made good progress on your study of the United Kingdom, you will be able to work on the questions on pages 48 to 49.

Scotland, Ireland, North-east England

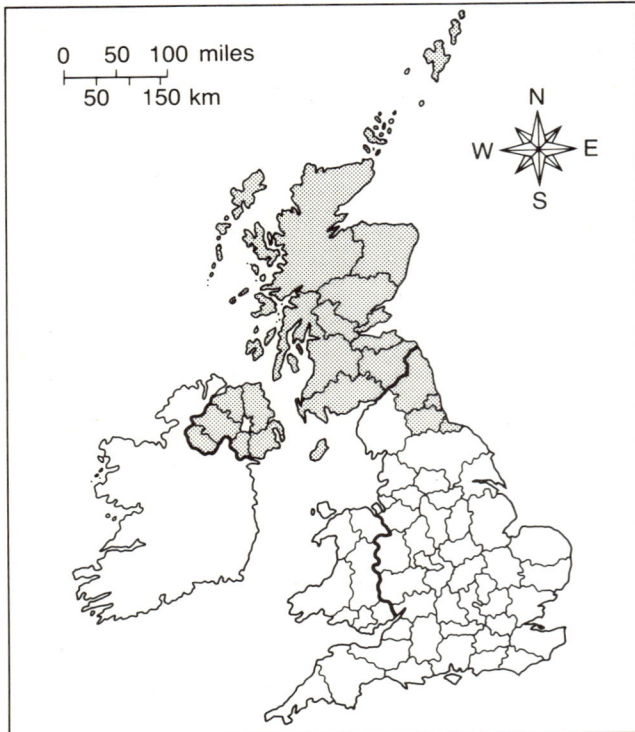

Fig 2.3 ● Scotland, Ireland and North-east England, shown in grey tone

Fig 2.4 ● Ban Foot Ferry, Lough Neagh, Co. Armagh

The map shows that Scotland, Ireland and north-east England together make up more than half the total land area of the British Isles. However, hardly more than one out of every six people in Britain lives in these areas. There are several big cities but also many large areas of open countryside.

Name	Population (1981) within administrative border	Not far from
Republic of Ireland (Eire)	**3 400 000**	Wales
Dublin	525 000	Wales
Northern Ireland (Ulster)	**1 550 000**	Scotland
Belfast	322 000	Lough Neagh
Scotland	**5 100 000**	Northern England
Edinburgh	450 000	Glasgow
Shetland Islands	23 000	Orkney Islands
Isle of Man	**65 000**	Mull of Kintyre

Fig 2.5 ● Populations and locations of some cities and areas

You will probably need to look up the answers to some of these questions. As you do so, you should be finding out about important places in these regions.

1. Which of the places listed in Fig 2.5 has the smallest population?
2. Which of the three capital cities listed has the largest population?
3. What is the name of the firth near Edinburgh?
4. What is the name of the firth near Glasgow?
5. What hills lie to the south of Inverness and to the west of Aberdeen?
6. For what sport is the Isle of Man famous?
7. Eire (the Republic of Ireland) is not part of the United Kingdom. Where does the Irish Parliament (Dail Eireann) meet?

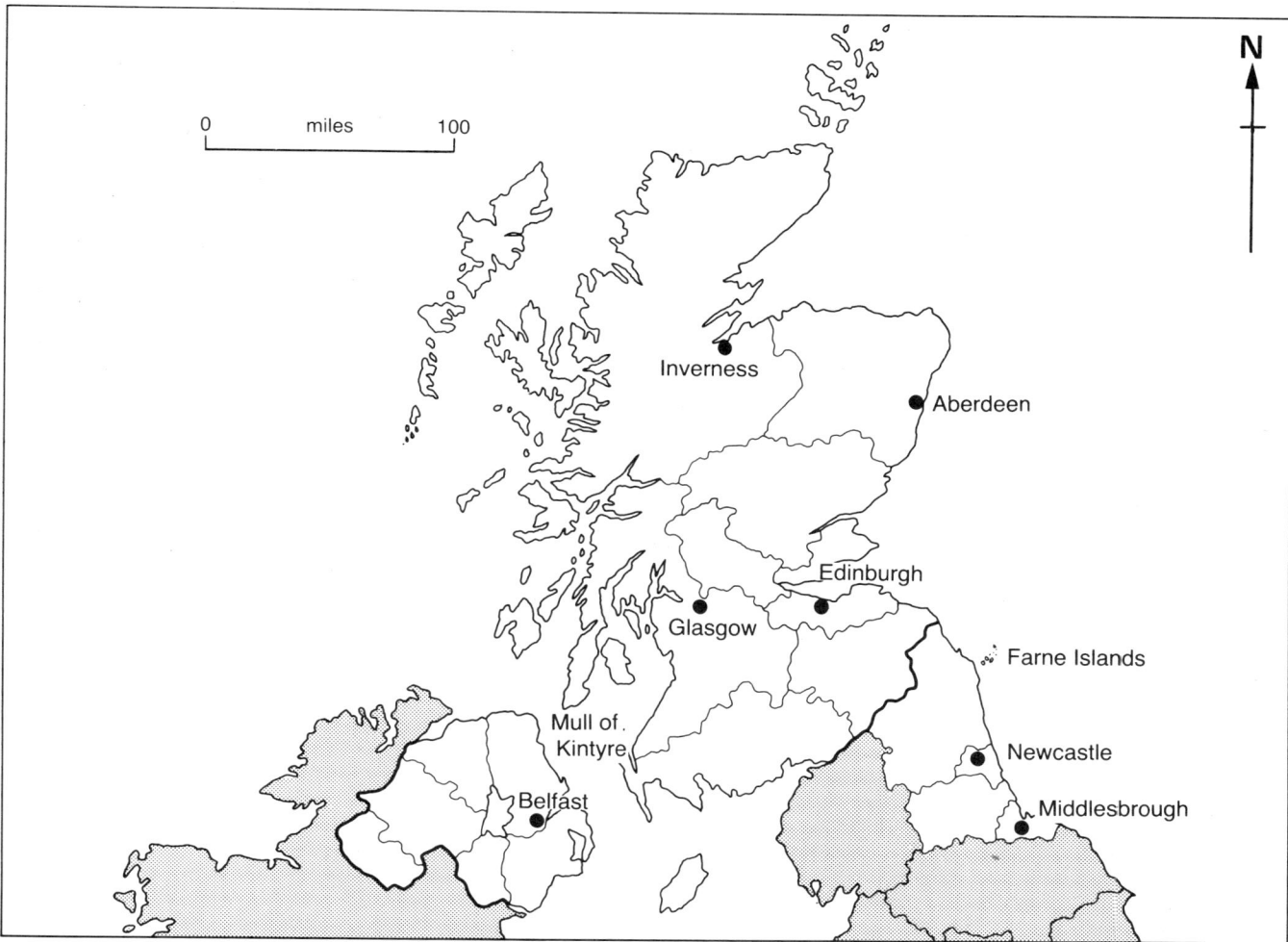

Fig 2.6 ● Detailed map of Scotland, Ireland and North-east England

8. How many counties are there in Northern Ireland (Ulster)?

9. How far is it to London from the Farne Islands off the coast of Northumberland?

10. From which port can you take a ferry to Belfast?

11. What are the names of the three provinces in the Republic of Ireland?

12. What is the name of the river that flows just north of Middlesbrough?

13. What is the name of the river that flows through Newcastle? In what county is Newcastle located?

14. What is the name of the sea lough near Belfast?

15. In which National Park can you see Hadrian's Wall? Which other National Park extends just into Cleveland?

16. What famous animal do some people keep looking for near Inverness?

17. What is the third largest city in Scotland, and a famous fishing centre?

18. Which city is on the River Tay, important both as a centre for offshore oil and for the electronics industry?

19. What industries have been very important in the growth of Glasgow?

20. What big town is just across the river from Middlesbrough?

Project

Scotland, Ireland, and north-east England all have landscapes with big hills and long coastlines. Build up a project that describes the main features of each part, with maps and photographs. Many of the rural areas have traditional crafts and festivals, which you could also describe. You might like to contrast these with the latest industrial developments in the areas.

London and the South-east

As the map (Fig 2.11) shows, the area around London and south-east England is only a small part of the British Isles. However, roughly a quarter of the whole population lives in this crowded corner.

Warmer, drier weather than in some other regions, combined with better chances for jobs, make this part attractive for many people.

Fig 2.7 ● Visitors to Kew Gardens experience the temperature, humidity and plants of the 'moist tropics'

Place	Population (1984) within administrative border	Near to
London	6 700 000	
Luton	164 000	Dunstable
Milton Keynes	107 000	Woburn Abbey
Brighton	146 000	Eastbourne
Southend	157 000	Thames estuary
Basildon	152 000	
Thurrock	127 000	River Thames
Reading	124 000	Thames valley

Fig 2.8 ● Populations and locations of some cities and towns

Airports and ports (1984 figures)

Heathrow	38 000 000 passengers to over 200 destinations
Gatwick	25 000 000 passengers to over 100 destinations
Stansted	8 000 000 passengers to various destinations
Dover	14 000 000 passengers to 4 destinations
Harwich	2 000 000 passengers to 4 destinations

Fig 2.9 ● Passengers using airports and ports

You may need to look up some of the answers to these questions about the most populated part of the British Isles.

1. Measured by the number of passengers a year, which of the airports and seaports in Fig. 2.9 is the most important?
2. Which of these is *not* in London? Houses of Parliament, Buckingham Palace, Windsor Castle, Westminster Abbey.
3. To how many different countries can you sail from Dover?
4. How many foreign ports are served by the ferries from Harwich?
5. What hills do you cross between London and Brighton?
6. Which is the narrowest point of the English Channel?
7. What is the name of the water between Kent and Essex?
8. Which three motorways might you use to drive from Dover to Milton Keynes?
9. Which of these phrases describes one of the main types of work in London? Farming, office work, making heavy machines, mining.

10. What do the City of London, Ealing, Hounslow and Tower Hamlets have in common? How many other places could be in the same list?

11. Which compass direction would you follow from Heathrow to Stansted?

12. London is the capital of the United Kingdom. What provinces are united in the Kingdom?

13. How many different Underground lines are there on London Transport?

14. Which of these parks is nearest to Piccadilly Circus in London? Hampstead Heath, Regent's Park, Bushy Park, Epping Forest?

15. Name the main-line rail stations in London. Which big cities can you reach from each station?

16. Which of these places is the oldest? London Zoo, the Tower of London, the Barbican Centre, the Science Museum?

17. Why was the Thames Barrier built?

18. Where is the Isle of Sheppey, and why do many people go there in summer?

19. For what products is the farmland in Kent famous?

20. Which motorway goes all the way around London?

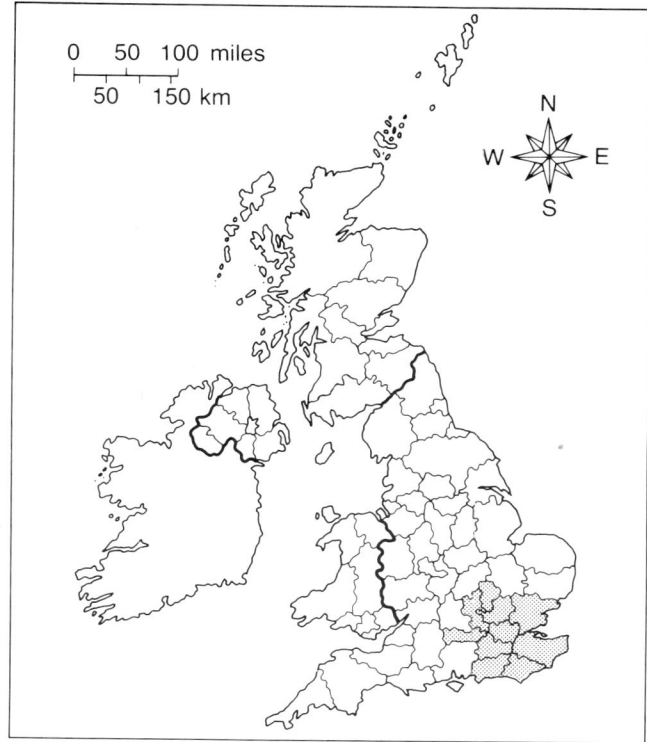

Fig 2.11 ● London and the South-east, shown in grey tone

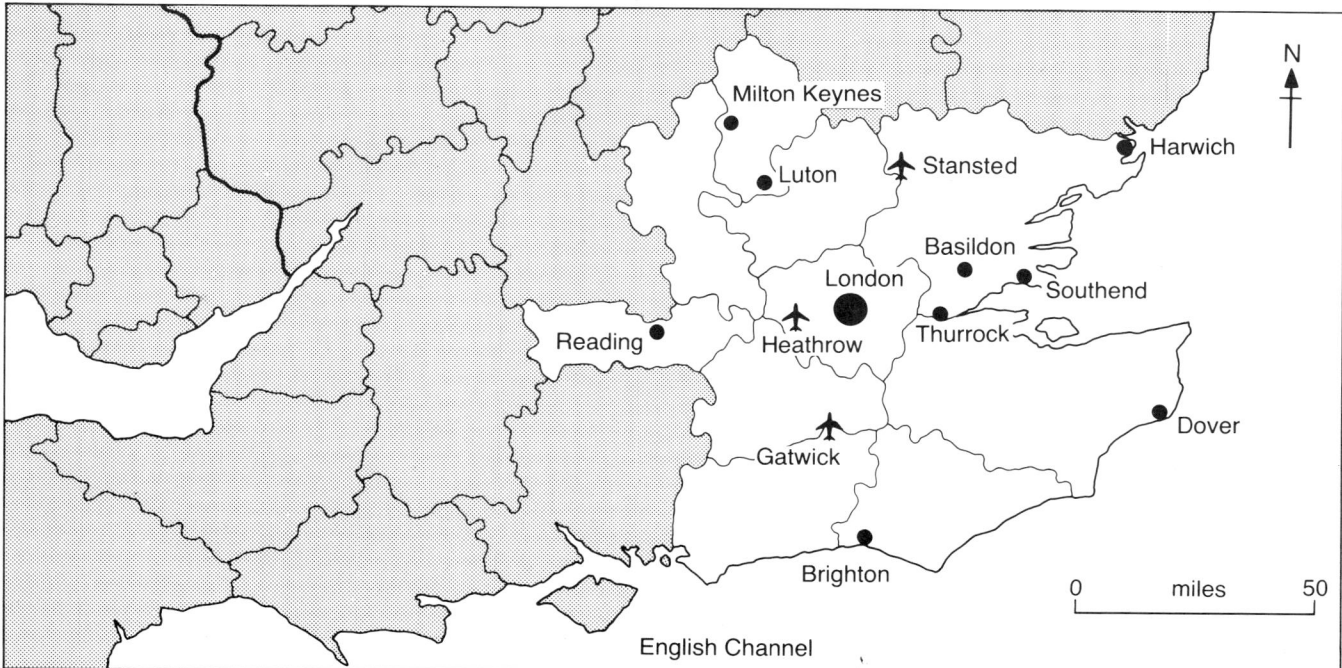

Fig 2.10 ● Detailed map of London and the South-east

Project

There are many places of interest to the tourist in the London area. Try to build up a guide to the ones that you would like to show a visitor if you had the chance.

West Midlands, North-west England, Mid and North Wales

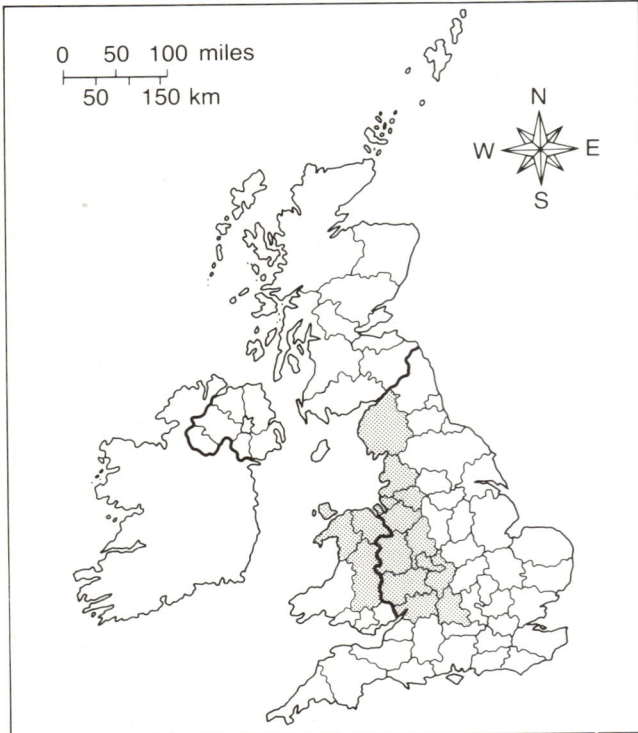

Fig 2.12 ● West Midlands, North-west England, Mid and North Wales, shown in grey tone

This area has a great deal of beautiful scenery. Almost all of it except the district around Liverpool and Manchester is hilly, and much of it is mountainous. The weather is often milder than in the east. There is usually rather more rain. Apart from around the big cities, much of the countryside is open grassland or woodland.

Place	Population (1984) within administrative border	Near to
Carlisle	72 000	Penrith
Blackpool	147 000	Preston
Liverpool	510 000	Birkenhead
Manchester	449 000	Bolton
Chester	58 000	Wrexham
Stoke-on-Trent	252 000	Crewe
Birmingham	920 000	Wolverhampton
Stratford-upon-Avon	21 000	Leamington Spa
Oxford	99 000	Cotswold Hills

Fig 2.14 ● Populations and locations of some cities

1. Which of the places in Fig 2.14 are especially popular with *British* holiday-makers?
2. Which of these places are especially popular with *foreign* tourists to Britain?
3. Which towns or cities in Fig. 2.16 are on the River Trent?
4. Which towns or cities in Fig. 2.16 are near the Peak District?
5. There are two National Parks wholly in this area, and two more that extend into it. Which ones are they? For what is each one famous?
6. Which motorway goes near Blackpool? What big cities does this motorway pass?
7. Which river flows between Liverpool and the Wirral? What is the county it runs through called?
8. What are the names of the towns in the potteries that are now included in Stoke-on-Trent?
9. What famous teams for popular sports come from around Birmingham, Liverpool and

Fig 2.13 ● The Bull Ring in Birmingham City centre

Manchester? What special sports facilities are there in these districts?

10. What two small towns does the River Avon pass after Stratford? Where does it join the River Severn?

11. What islands are north or north-west of Snowdonia?

12. Which of these lakes is longest? Ullswater, Derwent Water, Bassenthwaite Lake, Windermere?

13. How did ocean-going ships reach Manchester docks, more than 35 miles from the sea?

14. Which city has a world-famous University that is over six hundred years old?

15. Are there any high hills or large areas of open country in the West Midlands?

16. What big cities apart from Birmingham are in the West Midlands?

17. Which of the cities in this area has an airport that caters for more than five million passengers a year?

18. How high is Snowdon, and what ways are there of getting to the top?

19. How do drivers cross the River Mersey to go to and from Liverpool?

20. Near which of these cities is the National Exhibition Centre?

Fig 2.15 ● Bottle ovens used for firing pottery, Stoke-on-Trent

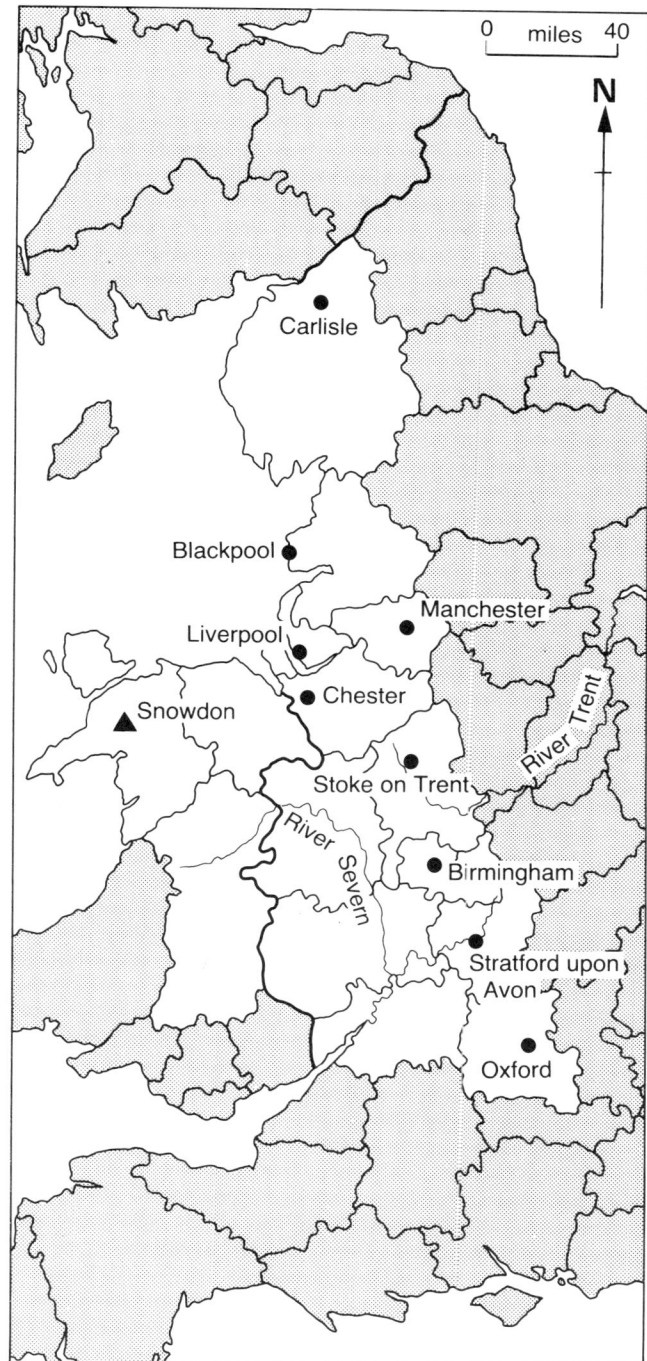

Fig 2.16 ● Detailed map of West Midlands, North-West England, Mid and North Wales

Project

There are many industries that have been linked with this area. Try to build up a project describing where and how these industries developed and what they were like. In what ways does the history of these industries interest visitors?

East Midlands, East Anglia, East Coast

This area has much good farm land and many of the country's best crops come from these parts. Coal is an important natural resource and there are also a few small oil wells.

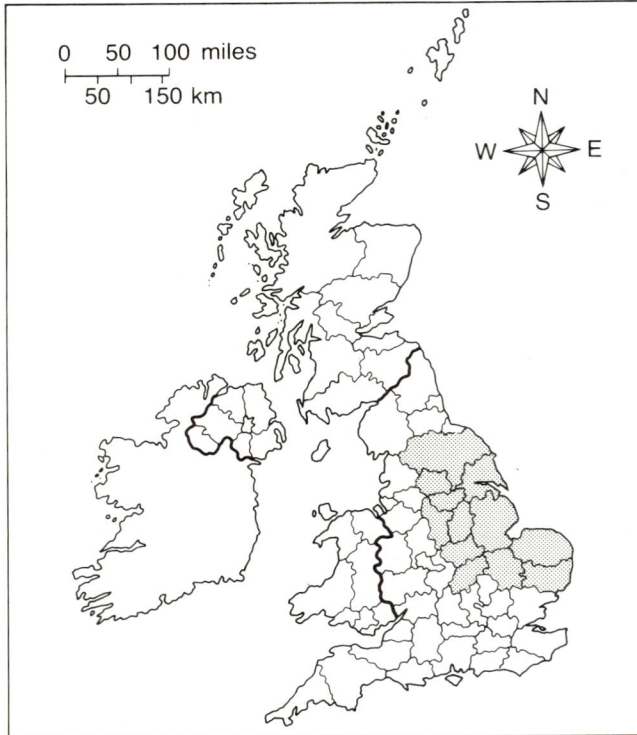

Fig 2.17 ● East Midlands, East Anglia and the East Coast, shown in grey tone

Fig 2.18 ● Wicken Fen, Cambridgeshire

Name	Population (1984) within administrative border	A centre for:	Main access road
York	100 000	Yorkshire moors	A64
Leeds	449 000	The Dales	M62/A1
Sheffield	447 000	Peak District	M1
Kingston-upon-Hull	268 000	Humber Estuary	M62
Nottingham	271 000	Sherwood Forest	M1
Derby	215 000	Vale of Trent	M1
Leicester	280 000	Charnwood Forest	M1
Cambridge	90 000	Fens	M11
Norwich	122 000	Norfolk Broads	A11
Ipswich	120 000	Orwell, Stour, Deben	A12
Felixstowe	21 000	North Sea	A12

Fig 2.19 ● Populations and locations of some cities

1. Which of the towns in Fig 2.19 are important seaports?
2. In which town is the Jorvik Viking Centre?
3. Which two of these cities are very close to each other?
4. Which city has an old and famous university?
5. Which of these cities are near coal mining industries or big power stations?
6. Which of these is an old city in a large agricultural district?
7. Which of these cities has, among its engineering factories, divisions concerned with Rolls-Royce and British Rail?
8. Which of these has, among its major industries, factories making bicycles and products for chemists' shelves?
9. Which very large city has varied work in offices, printing, clothing and engineering?
10. Why are the Pennine Hills often called 'the backbone of England'?
11. Which of these cities are on the River Trent?
12. Which of these cities is between Halifax, Wakefield and Harrogate? Which county is it in?

13. Which of these cities has been famous for high quality steel? Which county is it in?

14. Which city has big industries concerned with footwear, hosiery and knitwear?

15. The chart shows the unusually even spread of main roads leading into one of these cities. Which one is it?

16. Which of these cities is between Colchester and Lowestoft?

17. Which of these cities can be reached easily from the M1?

18. There are three National parks in or partly in this area. Which ones are they and for what is each one famous?

19. Which of these cities is furthest west?

20. Which of these is nearest to The Wash?

Fig 2.20 ● The Peak District

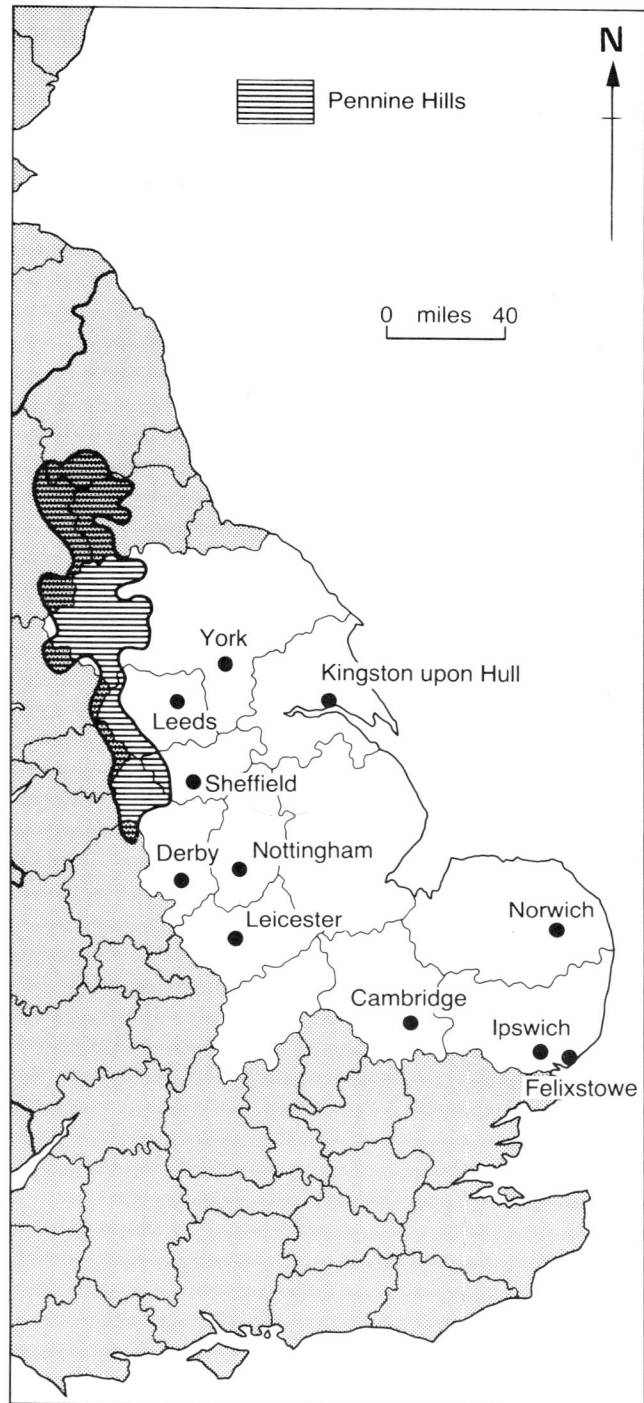

Fig 2.21 ● Detailed map of East Midlands, East Anglia and the East Coast

Project

There are many different kinds of landscape in this area. Try to find out what each one is like. What is the difference between a mere and a moor, a broad and a breck, a fen and a fell, a dale and a vale, a well and a wold? They are all to be found in this area, as well as in many others. Where are they, and what does each look like?

South and West Wales, West and South-west England

The western parts of England and Wales are largely green with rolling hills. They are clothed with woods and meadows that are kept a rich green by the warm moist air that often blows in off the Atlantic Ocean. There are relatively few big cities or major industrial areas. The distances between the big centres is usually greater than in other parts of England.

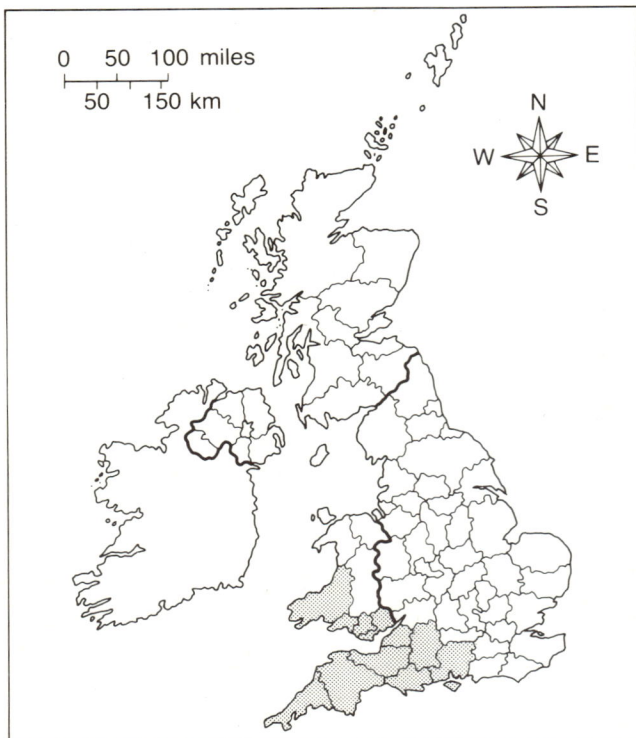

Fig 2.22 ● South and West Wales, West and South-west England shown in grey tone

Fig 2.24 ● Part of Bristol

Fig 2.23 ● Populations and locations of some cities

Name	Population (1984) within administrative border	Not far from	Next railway station East/North	Near motorway
Swindon	91 000	Wiltshire Downs	Didcot	M4
Southampton	204 000	South Downs	Winchester	M27
Portsmouth	179 000	South Downs	Petersfield	M27
Bournemouth	144 000	New Forest	Christchurch	A31
Bath	80 000	Cotswolds	Trowbridge	A4
Bristol	388 000	Mendips	Swindon	M4/M5
Cardiff	273 000	Rhondda	Newport	M4
Swansea	168 000	Fforest Fawr	Bridgend	M4
Torbay	116 000	Dartmoor	Exeter	A380
Plymouth	243 000	Bodmin Moor	Totnes	A38

1. Which cities in this table are near the Severn estuary?
2. Which cities in this list are near the English Channel?
3. What sports is the Isle of Wight famous for?
4. Which cities are on the River Avon?
5. What is the name of the county that includes the Pembrokeshire National Park?
6. Which is the nearest city to Land's End?
7. Which city is the capital of Wales?
8. What are the distinctive features of Exmoor and Dartmoor National Parks?
9. St Mary's is the largest of a group of 140 islands. Which islands are they?
10. Jersey is the largest of another group of islands. Name eight others in this group.
11. Which county has had tin mines for more than two thousand years?
12. How far to the south-west can you travel on the M5?
13. Which town grew very quickly during the 1980s?
14. Which two cities in this list are important naval bases?
15. Which big city between Weymouth and Southampton is a great holiday centre?
16. Which city includes the Gower Peninsula within its boundaries?
17. Which city has been a port for long-distance ocean journeys, and is also a container port?
18. Which city has hot springs, famous since Roman times?
19. Which city has Clifton Gorge, a suspension bridge and *SS Great Britain*?
20. What are the separate towns that have been joined into the district known as Torbay?

Project

The sea is important around Wales and South-west England. Find out the names of the three Channels that wash against these shores, where they become parts of the Irish Sea and Atlantic Ocean, and the names of the chief bays and headlands along these coasts. Which areas of the coastline are particularly famous and what are they like? What is the Severn Barrage and why do people want to have it built?

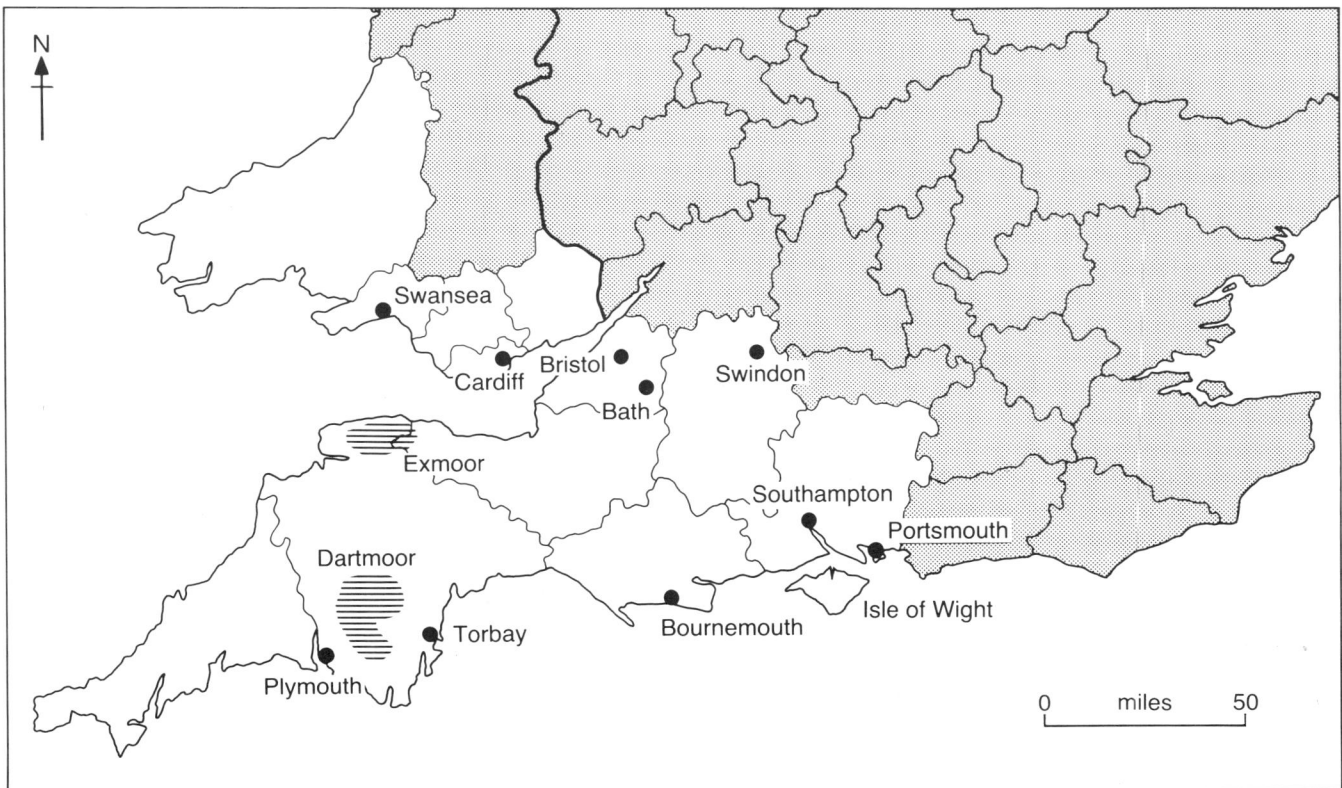

Fig 2.25 ● Detailed map of South and West Wales, West and South-west England

Long Distance Journeys

American Photographers' Tour

Some American visitors want to plan their own tour of the United Kingdom. They are particularly interested in the photographic challenges of Britain's rural areas. They have made their own list of the places where they want to take their photographs. These are given in the list below. From which of the Tourist Information Boards do they need details?

- Land's End
- Pembrokeshire National Park
- Snowdon
- Derwent Water
- The Loch Ness Monster
- The Cairngorms
- The Kent hopfields and oast houses
- The fens
- The Yorkshire moors
- The white cliffs of Dover

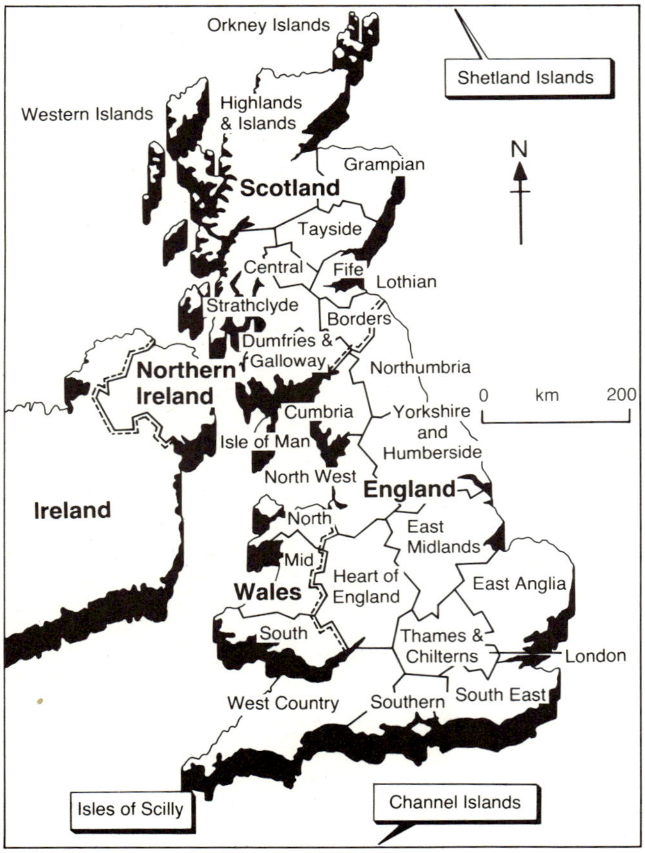

Fig 2.26 ● Tourist Information Boards

Plan the Route

A large importing company receives its goods at Kingston-upon-Hull, some by sea and some by air. Most of its lorries are fully booked on the normal schedule of calls, but urgent consignments need to be delivered to the following places. There are only three lorries available and between them they have to visit all the places. Which places would you ask each lorry to visit, and in what order would you put them to make the shortest journeys?

- Aberdeen (Ab)
- Blackpool (Bl)
- Dover (Do)
- Gatwick airport (Ga)
- London (Lo)
- Middlesbrough (Mi)
- Norwich (No)
- Oxford (Ox)
- Swansea (Sw)
- Birmingham airport (Bi)
- Brighton (Br)
- Dundee (Du)
- Glasgow (Gl)
- Luton (Lu)
- Newcastle upon Tyne (Ne)
- Nottingham (Not)
- Sheffield (Sh)
- Swindon (Swi)

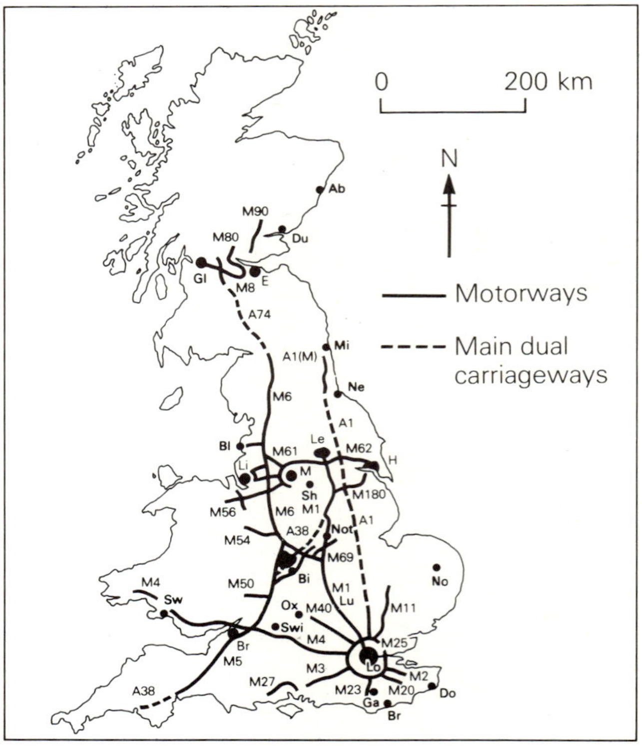

Fig 2.27 ● Motorways in Britain

Avoid the Delays

You are working for a television company based in central London. Your job is to see that all the camera crews can manage to reach their assignments with plenty of time to set up their equipment before the reporters arrive. You have just been given a list of motorway maintenance work. This morning in May the crews are setting off on the following assignments. Which ones do you need to warn about possible delays? Where might they expect to be held up?

Assignments

- Filming yacht races around the Isle of Wight
- Filming a national conference in Blackpool
- Filming an international trade fair in Sheffield
- Filming holiday traffic near Bristol
- Filming Welsh Eisteddfod events near Cardiff
- Filming major developments at Manchester airport
- Filming the volume of traffic passing through Liverpool

A friend in Leicester phones you at work because he wants your advice about the best route to take to Southampton. In the light of the motorway maintenance going on, what advice would you give?

When you have finished work today you want to go off to Nottingham for the weekend. What delays might you experience?

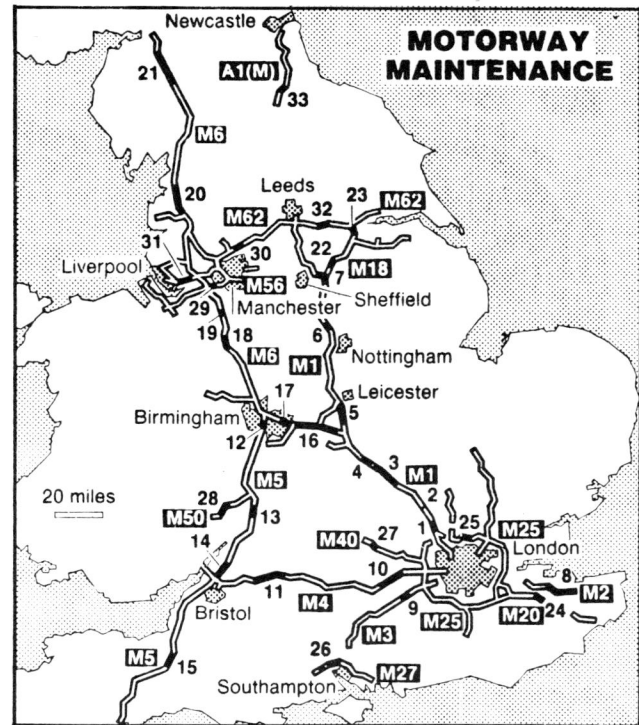

Motorway repairs in 1986-7 (with map references in bold):

M1: **1**, May-December; **2**, April-October; **3**, September-December; **4**, March-September; **5**, May-June; **6**, September-October; **7**, August-October.
M2: **8**, September-December.
M3: **9**, May-July.
M4: **10**, May-August; **11**, April-July.
M5: **12**, June-September; **13**, April-August; **14**, April-June; **15**, September-October.
M6: **16**, May-June; **17**, July-September; **18**, April-May; **19**, September-October; **20**, April-October; **21**, July-September.
M18: **22**, August-October; **23**, June-November.

M20: **24**, May-November.
M25: **25**, May-November.
M27: **26**, April-June.
M40: **27**, ?.
M50: **28**, April-August.
M56: **29**, April-July.
M62: **30**, May-August; **31**, June-September; **32**, August-October.
A1(M): **33**, May-July.

Fig 2.29 ● Motorway repairs (from *The Times*)

Fig 2.28 ● M4 motorway leading towards London

Quiz

Spot the Errors

What is not correct in the following sentences?

1. York was named by homesick Americans who came from New York.
2. Bath, Blackpool, Brighton and Bournemouth are all in the south of England.
3. Bournemouth, Portsmouth and Plymouth are the only big towns on the south coast that are on the mouth of a river.
4. A county council makes decisions about roads, schools, the police, fire services and the armed forces in its area.
5. There are high hills in every county in the north-west – Shropshire, Staffordshire, Lancashire, Merseyside, Greater Manchester and Cumbria.
6. Every town on the east coast is further east than London.
7. There is a motorway that goes all the way from London to Swansea and another from Bristol to Plymouth.
8. The motorways out of London are the M1, M2, M3, M4 and M5.

Clever Clogs

1. Which town lies due north of Blackpool?
2. How many counties are there in England and Wales?
3. How many people live in the whole of Ireland?
4. Can you explain why the number of passengers through the British ports each year is much greater than the whole population of the United Kingdom?
5. Which natural resources are very important to Britain's prosperity and where are they found?
6. Which counties of the United Kingdom are partly below sea level?
7. Why is the climate mild and moist in the south-west of England?
8. Why are there more tourist boards in the north of the British Isles than in the south?

Picture Search

● Where is this landscape?

● In what part of the United Kingdom would you find this ferry?

● Is this building in the West Midlands, North Wales or East Anglia?

Pick the Word

1. Which of these do you link with Swansea? Tennis / ice-skating / rugby / horse racing
2. Which of these do you link most with the Romans? Dundee / Bath / Middlesbrough / Dublin
3. Which of these is part of the British Isles but not of the United Kingdom? Isle of Man / Eire / Orkney / Channel Islands
4. Which of these is furthest north? Belfast / Glasgow / Dublin / Shetland
5. The biggest city in the north-west is Birmingham / Liverpool / Manchester / Stoke-on-Trent
6. An important port on the east coast is Hull / Scarborough / Skegness / The Wash
7. There are very few trees in the Cotswolds / the Mendips / Fforest Fawr / Dartmoor
8. A big city that does not have full motorway connections with London is Bristol / Birmingham / Liverpool / Newcastle upon Tyne

Crossword

Make a tidy copy of this grid and then solve the clues. You will find that many of the important words were used in this book between pages 33 and 47.

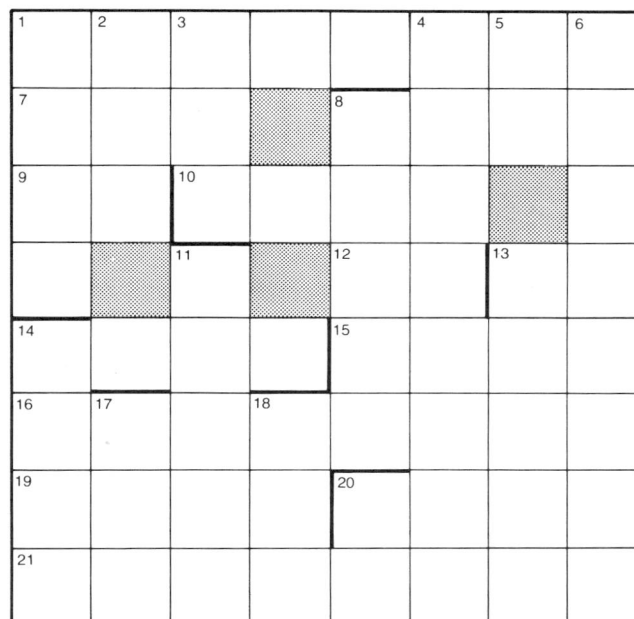

Clues across:

1. These hills are often called the 'backbone of England' (8)
7. How many capital cities are there in England? (3)
8. The Scottish name for a lake (4)
9. The way 'road' is written on an address (2)
10. A waterfall built by man to control the flow of a river (4)
12. Short for 'street' (2)
13. Middle two letters of 'Bootle' (2)
14. Found in the sea, and mined in Cheshire (4)
15. These four letters do not make a word. They are the initials of the rivers Trent, Humber, Conway and Lee (4)
16. The water that lies between Wales and Ireland (5,3)
19. You may need to know the number of this to find a particular map in an Atlas (4)
20. Name of a forest in Gloucestershire and Gwent where opencast coal is found (4)
21. The northern part of the United Kingdom (8)

Clues down:

1. Dover is important because it is one of these towns (4)
2. The furthest part of Cornwall is this bit of land (3)
3. A common part of a place name, may be with the word 'castle' as well (3)
4. The water that lies between the east of England and Denmark (4,3)
5. The initials of the words 'European Community' (2)
6. The islands that are the northern limit of the United Kingdom (8)
8. Names or facts put in a column (4)
11. A county in the north west of the Republic of Ireland (5)
13. A large expanse of water, such as that between Ireland and America (5)
14. Takes little drinks (4)
17. One of the motoring organisations that helps when you break down (3)
18. What the sun does in the west (3)
20. Short for 'decilitre' (2)

World Regions

There are over 150 different countries in the world, and it can be hard to remember where each one is. People find it helpful therefore to remember first of all the names of particular regions of the world. If you hear about Burundi, and are told that Burundi is in Africa, you are well on the way to finding it quickly on any map.

Ten regions of the world to know about are:

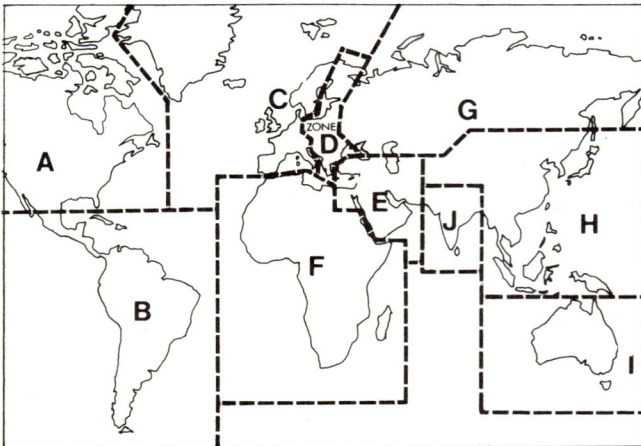

Fig 3.1 ● Regions of the world

Region A:	**North America**
Region B:	**Central and South America**
Region C:	**Western Europe**
Region D:	**Eastern Europe**
Region E:	**Middle East**
Region F:	**Africa**
Region G:	**The USSR**
Region H:	**The Far East**
Region I:	**Australia and New Zealand**
Region J:	**The Indian sub-continent**

Can you identify which of these regions each of these news items refers to?

1. An expedition is to be made to study the insects and other wildlife found at different heights on the steep cliffs of the Grand Canyon.
2. There are still doubts about the safety of Aborigines crossing the Gibson desert many years after nuclear tests were carried out there.

3. The water that flows down the Nile each spring is less valuable for irrigation as rich silts are held up in the Aswan Dam.
4. Food production in India has improved so much in recent years that the country is now able to export grain to others in need.
5. New snowfalls in the Alps have made skiing conditions the best so far this year.
6. The weather in Leningrad and Moscow has been particularly cold recently. The daytime temperatures have remained below −30°C.
7. Eruptions of the volcano Paricutin have prompted scientists in Mexico to watch more closely for any possible eruption of the giant old volcano Popocatepetl.
8. Discussions continue between the Greek and Turkish communities in Cyprus.
9. The population of Alma-Ata, the capital of Kazakhstan SSR, has now passed one million as the region continues to prosper.
10. There are a total of 7090 islands in the Philippines, and the population is almost the same as that of the United Kingdom.

International Pilot

What place names are missing here?

Kersti is an international airline pilot. Last week she flew directly to Nigeria and then east across the continent of (1)____ to land at Cairo in (2)____. Next she touched down in Jerusalem in (3)____ and finally at Warsaw, the capital of (4)____, before returning to Gatwick

Next week she is scheduled to visit countries on the west of the (5)____ Ocean, starting at Ottawa, the Federal capital of (6)____. She will then fly south over (7)____ to Brazil. Here she will have good views of the great river (8)____ before touching down at the capital of Argentina, (9)____, and flying back across the (10)____.

Some Key Countries

Here are a few notes about some of the world's countries. In which region is each?

Canada, a member of the Commonwealth, is a federal state with twelve provinces. About two thirds of the 24 million people speak English as their main language, and about one third speak French. About two million Canadians have a language other than these as their native tongue.

The **United States of America** is a federation of fifty states and has almost ten times the population of Canada. The USA is one of the world's richest countries and is also very powerful in world affairs.

Brazil is one of the largest countries in the world, and has vast areas of forest and good land that are not yet fully developed. It has rapidly developing industries in vehicles, ships and other manufactured goods. Important agricultural products include soya beans and coffee. The great and prosperous city of São Paulo is growing rapidly but like many third world cities does have shanty towns. In 1960, the Government set up a new capital city, Brasilia, five hundred miles inland. Almost half of the people in the country are under the age of 15.

Fig 3.2 ● A favela (shanty town) in Rio de Janeiro, Brazil

Argentina is another large South American country which is also developing very rapidly. It is 2300 miles (3700 km) from north to south and over 900 miles (1400 km) wide. However, the population is only just over half that of the United Kingdom.

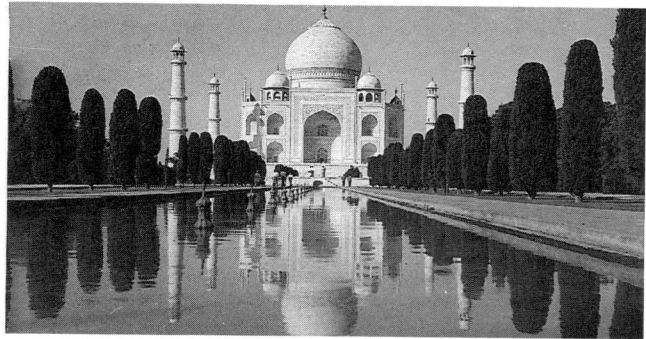

Fig 3.3 ● What is this famous landmark and in what country is it found?

Poland has long had close ties with the United Kingdom, and many thousands of British people have Polish ancestry. The country is now part of the Warsaw Pact and it is difficult for Polish citizens to visit Britain. Considerable trade continues between the two countries.

East Germany is the part of Germany that was administered by the USSR after the collapse of the Nazi government in 1945. Germany's former capital city, Berlin, is within Eastern Germany and is itself divided into Western and Eastern sectors. Citizens of East Berlin and East Germany find it difficult to obtain permission to visit the West.

Israel is a small country on the eastern shore of the Mediterranean. There is considerable trade with the United Kingdom, particularly Israel's exports of fruit.

Egypt is almost entirely desert, with only 3% of the land being cultivated. Most of the population, roughly the same number as in Britain, live in the north in or around Cairo. Further south, people can live only near the River Nile and many are still very poor. Many splendid monuments from the past, especially the Pyramids, attract large numbers of tourists.

Nigeria is one of the largest African countries and has more people than Britain. There are important oil fields, which provide much of the country's wealth.

South Africa is a large country with many high mountains and much excellent agricultural land. It is rich in gold and diamonds, metals and other minerals. There are about six million white and about twenty-five million black and coloured people. South Africa was a member of the Commonwealth until 1961, when it withdrew rather than alter its policy of Apartheid (separation of the races).

The USSR is a federation of fifteen states and a total of over 270 million people. The large cities of Moscow and Leningrad are those most often visited by tourists from the West.

China has about twenty citizens for every citizen in the United Kingdom. Rapid changes are taking place throughout the country, and it is becoming much easier for tourists from the West to travel around. One attraction is the Great Wall built over two thousand years ago and extending for over a thousand miles along the country's border.

Japan has become one of the world's richer countries since the Second World War. Britain's imports from Japan in 1984 were worth over £75 for every man, woman and child in Britain. Many of Britain's big firms have increasing links with Japanese firms.

Hong Kong (not an independent country) is a group of 236 islands plus an area of mainland bordering south-east China. Hong Kong Island is the commercial and business centre. The New Territories (which includes the mainland and rest of the islands) was leased to Britain for 99 years. From the year 1997, all of Hong Kong becomes a region of the People's Republic of China.

Australia is a member of the Commonwealth. Much of its great continent is desert, and the majority of the population lives along the eastern coast. Most of the country's exports are foods and minerals.

New Zealand is another member of the Commonwealth. With only just over three million people, its population is hardly larger than that of Wales. The main exports to Britain are butter and meat.

India is the largest member of the Commonwealth. There are more people in India than in all the rest of the Commonwealth combined, including the United Kingdom. India is a federal country with 22 states and nine Union Territories. The capital is Delhi.

Pakistan left the Commonwealth in 1972. It is to the north-west of India with varied land, from Himalayan Mountains in the north to areas of desert in the south. Main exports include raw cotton, fabrics, yarn and thread.

Bangladesh became a separate nation in 1971. It lies along the River Ganges, to the north-east of India. Bangladesh has about twice as many people as the United Kingdom, and is a member of the Commonwealth. The country is one of the world's poorest nations. Jute is a main export.

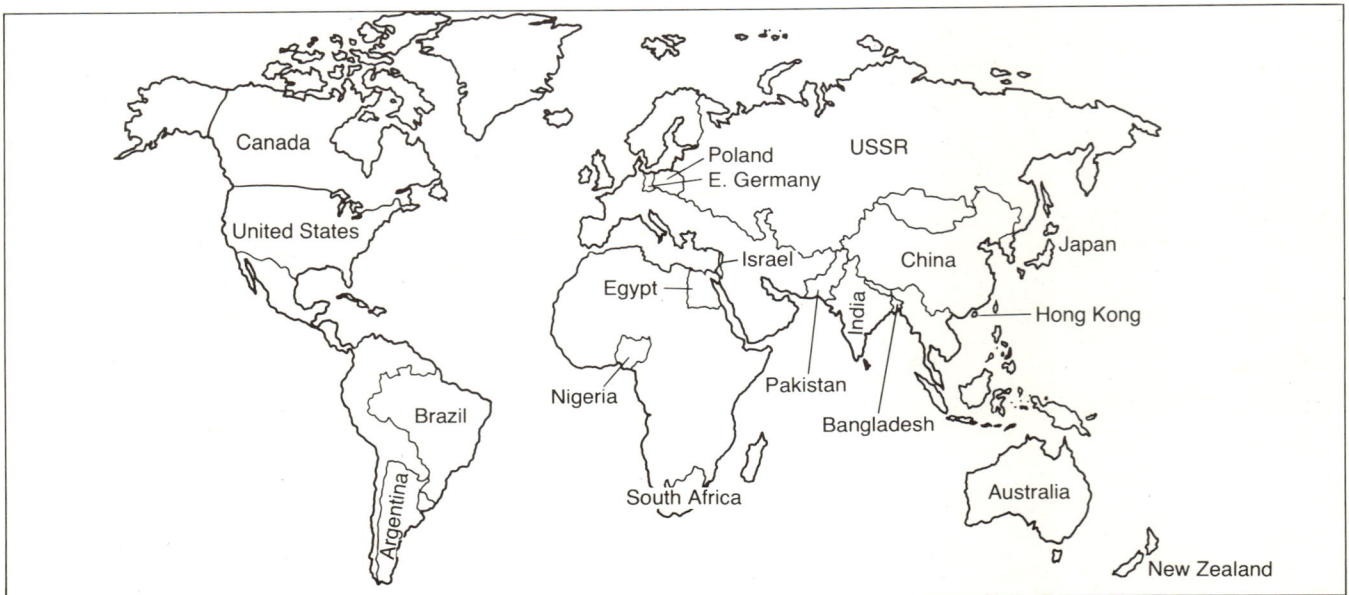

Fig 3.4 ● Some countries of the world

Three Rivers

Here are a few notes about some of the world's main rivers. In which region is each?

The Nile is 4145 miles (6632 km) long, flowing from the Ethiopian Highlands north through Egypt to the Mediterranean Sea. The Aswam Dam was built to help generate electricity and ensure a constant supply of water. It has also meant that the annual floods are now controlled, and a strip of land between five and fifteen miles wide is regularly cultivated along each bank of the river. However, the amount of water flowing down the river to the sea is not nearly as much as in many of the world's big rivers.

The Amazon is 4007 miles (6411 km) long and discharges a hundred times as much water into the Atlantic Ocean as does the River Severn. The Amazon's broad deep waters can be navigated upstream for 2300 miles.

The Congo (or **Zaire**) is 2920 miles (4672 km) long. It is the world's second biggest river, measured by the amount of water reaching the sea. It flows west through Africa to the Atlantic Ocean.

Fig 3.5 ● A felucca (small boat) on the River Nile

Fig 3.6 ● House on the Amazon River, Brazil

Cities of the World

Here are a few notes about some of the world's main cities. In which world region is each?

Fig 3.7 ● Peking (Beijing), China

Buenos Aires

This large modern city is the capital of Argentina. Buenos Aires is built on the banks of the great Parana river.

Moscow

Moscow is the capital of the huge Union of Soviet Socialist Republics (USSR) which stretches for thousands of kilometres over the Eurasian continent. Around the government headquarters, the Kremlin, are old buildings with domes and ornate decoration. There is also a new city with modern parks, blocks of flats and shopping centres. Theatres, museums and art galleries are among the things to see.

New York

The Manhattan skyline is a famous feature of New York, including the tall Empire State Building. There is also Central Park, the Broadway theatres and the shops of Fifth Avenue.

Peking (Beijing)

This is the capital city of China, one of the largest nations in the world. In Beijing, you can see the old Imperial Palace, known as the Forbidden City, and also the Summer Palace and the Temple of Heaven.

Singapore

Singapore is built on a tiny island at the tip of the Malay peninsula. In the city, you can see Chinatown and Arab Street, as well as Indian temples.

Tokyo

Tokyo is on the east coast of Japan, facing the Pacific Ocean. It is one of the world's most crowded cities. Japan is a country that extends for over a thousand miles over several islands and Tokyo is not far from the midpoint of these.

Largest Cities of the World

The following table gives the approximate populations of the ten largest cities of the world in the mid-1980s. New York and London are shown as well.

Name	Population (1984)	Rank order
Mexico City	16 000 000	1
Shanghai	12 000 000	2
Tokyo	12 000 000	3
Cairo	11 500 000	4
Paris	10 000 000	5
Tehran	10 000 000	6
Buenos Aires	10 000 000	7
Beijing (Peking)	9 000 000	8
Calcutta	9 000 000	9
Moscow	9 000 000	10
New York	7 000 000	15
London	7 000 000	17

Fig 3.8 ● The world's largest cities by population (1984)

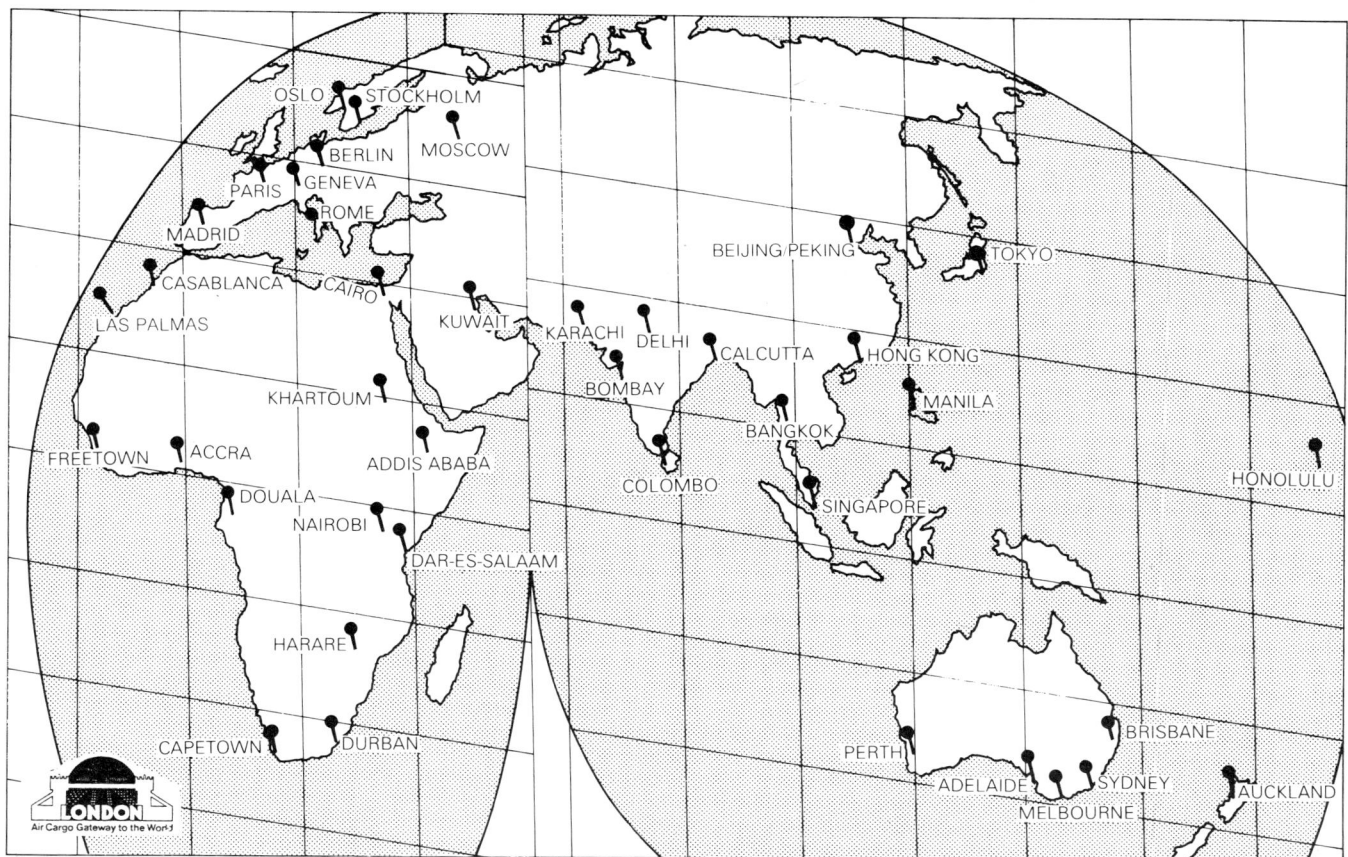

Fig 3.9 ● Some of the airports to which air freight can be flown directly from Heathrow

Use this map to work out what places a plane would be likely to pass on a direct flight from Heathrow to any of the big cities named.

If you won a competition that would allow you to fly to any far-off place, which one of these would you choose? What would you fly over on your way?

The World Calling

What place names belong in these gaps?

Roger is a telephone receptionist in a large international company. He has to work shifts because of the world's time zones. During the night shift he often receives calls made by people at work in Melbourne in (1)____ and Auckland in (2)____. Calls come in very early from Shanghai in (3)____ and a manufacturing centre in the Far East (4)____. A little later he receives calls from Dacca in (5)____, Bombay in (6)____, and Karachi in (7)____. Calls from Leningrad, (8)____, seem quite local, and may come in Britain's usual office hours.

The following airports are marked on the map. Which countries are they in?

- Accra
- Addis Ababa
- Bombay
- Cairo
- Manila
- Sydney

- Moscow
- Rome
- Casablanca
- Oslo
- Nairobi
- Auckland

It Can Only Be . . .

Can you match these descriptions?

1. An island city on the Equator
2. A large city on a small island north of the Equator
3. The city with the largest population in the world
4. A city of tall skyscrapers
5. The largest city in South America
6. City on the Nile
7. Capital of the country with the largest land area in Europe and Asia
8. City with the Temple of Heaven
9. On the great Parana River

Mountains and Oceans

Two thirds of the earth's surface is covered with water. The two great expanses are the Pacific Ocean and the Atlantic Ocean. The land is broken into the great continents of North and South America, Africa, Australasia, Antarctica, Europe and Asia.

Each day the earth spins once around its axis, which runs from the North Pole to the South Pole. The Equator is an imaginary line that marks the mid-point between the poles. On each side of the Equator are imaginary lines called the Tropics. These show the limits of the area where the sun moves directly overhead at midday with the varying seasons. Near the Poles are the imaginary lines, the Arctic and Antarctic Circles. These mark the limits of the area where the sun stays above the horizon all night at least once a year, according to the season.

The world has a number of great mountain ranges. The Himalayas (along the northern border of Pakistan, India and Bangladesh) have the highest peaks in the world, including Mount Everest. The Rockies and the Andes form a great ridge running almost the length of North and South America, also including many high peaks. In South America, several of the high peaks are still active volcanoes.

The world has a number of great rivers as well. The Amazon drains from the Andes mountains through the jungles of South America. The Congo (Zaire) drains from central Africa, also running through dense jungle. The Nile runs from the Ethiopian mountains through the edge of the Sahara desert, before finally flowing into the Mediterranean Sea.

Fig 3.10 ● The Grand Canyon, carved out by the Colorado River in South-west USA

Fig 3.11 • Sahara Desert

The world's greatest desert is the Sahara. In this vast area, far bigger than the whole of Europe, rain falls regularly only in mountainous areas. Rainfall is sparse elsewhere and shortage of water is a problem. Plant life outside the oases is mainly just thorny shrubs and coarse grasses.

World Physical Features

1. If you made a trans-Atlantic crossing from the United Kingdom, where might you end your journey?
2. Why is Christmas weather so different in countries south of the Equator?
3. Which country would you pass as your ship travels from the Atlantic into the Pacific?
4. How much difference in the length of daylight would you notice from summer to winter at the Equator?
5. From which country might you begin your journey through the Indian Ocean?
6. If you wanted to stand on your own shadow at midday at least once a year, where would you go?
7. If you wanted to see the sun shine all day and all night in June, where would you go?
8. If you wanted to see the sunshine all day and all night in December, where would you go?
9. How far is it from the Atlas Mountains at the north of the Sahara Desert to Lake Chad, at the south? How far is it from Mauritania in the west of the desert to Somalia in the east?
10. Which mountains would you fly over on a journey by air from New Zealand to Chicago (USA)?
11. How many days would it take a ship to go from Wellington (New Zealand) to Panama (South America)?
12. Which great river would you fly over on an air journey from London to Luanda (Angola)?
13. Which mountain range would you fly over on an air journey from Amsterdam (Netherlands) to Santiago (Chile, South America)?
14. How many days would it take a ship to go from Cape Town (South Africa) to Southampton if it can cover 400 miles a day?
15. Which mountain range would you fly over on a journey from Moscow (USSR) to Calcutta (India)?
16. Which two great rivers flow into the Atlantic Ocean, both with their estuaries near the Equator?
17. Which great river flows north into the Mediterranean Sea?
18. Why are there no major shipping routes between South America (Brazil, Uruguay or Argentina) and Africa?
19. Between which lines on a world map would you always expect to be warm?
20. In which areas would you expect to see icebergs?

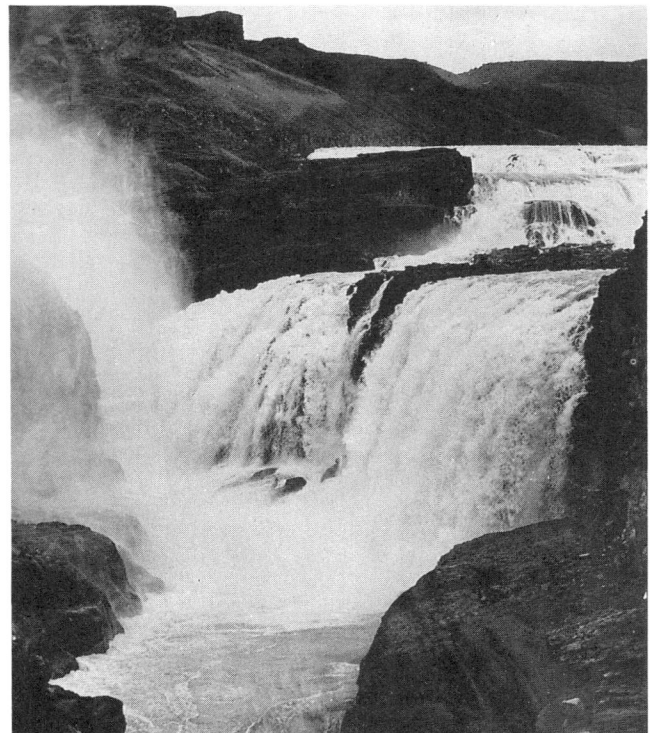

Fig 3.12 • Waterfall in Iceland

Countries of Europe

Fig 3.13 ● Dyke in The Netherlands

Country	Population (1984)	Area (sq m)	Exports to UK (£m)	Imports from UK (£m)	Capital city
Austria	7 552 000	32 000	593	321	Vienna
Belgium	9 863 000	12 000	3 692	3 051	Brussels
Denmark	5 116 000	17 000	1 660	1 197	Copenhagen
Eire	3 400 000	27 000	2 635	3 393	Dublin
Finland	4 844 000	130 000	1 249	684	Helsinki
France	54 832 000	213 000	5 885	7 082	Paris
West Germany	61 049 000	96 000	11 090	7 458	Bonn
Greece	9 740 000	51 000	279	354	Athens
Iceland	240 000	40 000	86	64	Reykjavik
Italy	57 080 000	131 000	3 814	2 902	Rome
Luxembourg	366 000	1 000	64	35	Luxembourg
The Netherlands	14 395 000	13 000	6 147	6 128	Amsterdam
Norway	4 146 000	149 000	3 852	968	Oslo
Portugal	10 031 000	34 000	645	386	Lisbon
Spain	37 834 000	197 000	1 668	1 332	Madrid
Sweden	8 331 000	173 000	2 416	2 889	Stockholm
Switzerland	6 482 000	16 000	2 490	1 549	Berne
Yugoslavia	22 800 000	99 000	108	163	Belgrade

Fig 3.14 ● Countries of Europe

1. Through which countries does the River Rhine flow?
2. In which countries are the Alps?
3. Which are the five largest countries in this list, in terms of population?
4. Which are the five smallest countries in this list, in terms of land area?
5. Which of these countries have land that lies north of the Arctic Circle?
6. Which of these countries have land that lies south of the 40 degree line of latitude?
7. Which of these countries are members of the EEC? What does it mean if they are?
8. Which five of these countries export most to the United Kingdom?
9. Which five of these countries receive most exports from the United Kingdom?
10. Which of these countries receive more exports from the UK than are sent to the UK?
11. What are the main products that the UK buys from other European countries?
12. What are the main products the UK sells to other European countries?
13. Are there any other countries in Europe not shown in this list?

Fig 3.16 • Olive trees on the Greek island of Crete. Olives and olive oil are important exports

Fig 3.15 • A vineyard in France. Wine is an important export

Country Clues

Which of the countries in the list on page 58 matches these descriptions?

1. Largely uninhabited except around the coast.
2. No coastline.
3. Much of the land is below sea level.
4. Capital is on an island.
5. Largest land area in western Europe.
6. Mediterranean Sea on west and east coast.
7. Has the highest mountains in Europe.
8. Lies between The Netherlands and France.
9. Lies west of Spain.
10. Has a long border with the USSR.

Famous Cities of Europe

The cities in list A below are some of the most famous of Europe. List B gives a few clues about why they are especially famous.

Can you match the clues to the right cities?

List A

- Amsterdam
- Athens
- Berlin
- Brussels
- Cologne
- Copenhagen
- Florence
- Madrid
- Paris
- Prague
- Rome
- Salzburg
- Venice
- Vienna
- Zurich

List B

- 1000 bridges, lively Dutch metropolis
- Centre for classical music, Boys' Choir, Orchestras
- Capital of Belgium, international centre with EEC administration
- Home of democracy 2500 years ago
- Bustling capital of Denmark
- Former capital of Germany, now divided by the Wall into Eastern and Western sectors
- Beautiful buildings, Rhine valley
- Largest Swiss city
- Art centre, Petrarch, Dante, da Vinci, Michelangelo
- Capital of France, Eiffel Tower, Champs-Élysées
- Sunny capital of Spain, fine old buildings, Flamenco dancing
- Capital of Czechoslovakia, with forests and mountains around
- Beautiful traditional Austrian city
- Ancient capital city of the Roman Empire
- Canals, medieval buildings

If you could visit just one city from this list, which one would you visit if you were interested in seeing:

1. Canals
2. The Eiffel Tower
3. Historical architecture from 2500 years ago
4. The former capital of Germany, and the effect of partition on the country
5. Flamenco dancing
6. Many famous orchestras and a world famous boys' choir
7. Beautiful paintings by Michelangelo
8. The Rhine valley
9. A city with 1000 bridges, with tulip bulbfields nearby

Which city would you yourself choose to go to? Why?

Fig 3.17 ● The Colosseum in Rome, built 72–80AD

Look at the list of cities below. In which country is each found? Which are capital cities? What European capitals are not on the list?

Name	Population (1984)	Approximate distance from London (miles)
Athens	3 027 000	1 450
Brussels	1 000 000	200
Copenhagen	575 000	600
Hamburg	1 618 000	450
Luxembourg	79 000	300
Madrid	3 188 000	780
Oslo	449 000	750
Paris	8 707 000	200
Rome	2 827 000	900
Rotterdam	559 000	180

Fig 3.18 ● Some cities in Europe (1984)

HOLIDAY WEATHER

■	−5 to 0C / 23 to 32F	▣	15 to 20C / 59 to 68F
□	0 to 5C / 32 to 41F	▨	20 to 25C / 68 to 77F
▢	5 to 10C / 41 to 50F	▩	above 25C / above 77F
▦	10 to 15C / 50 to 59F		

1000 km

0   500 miles

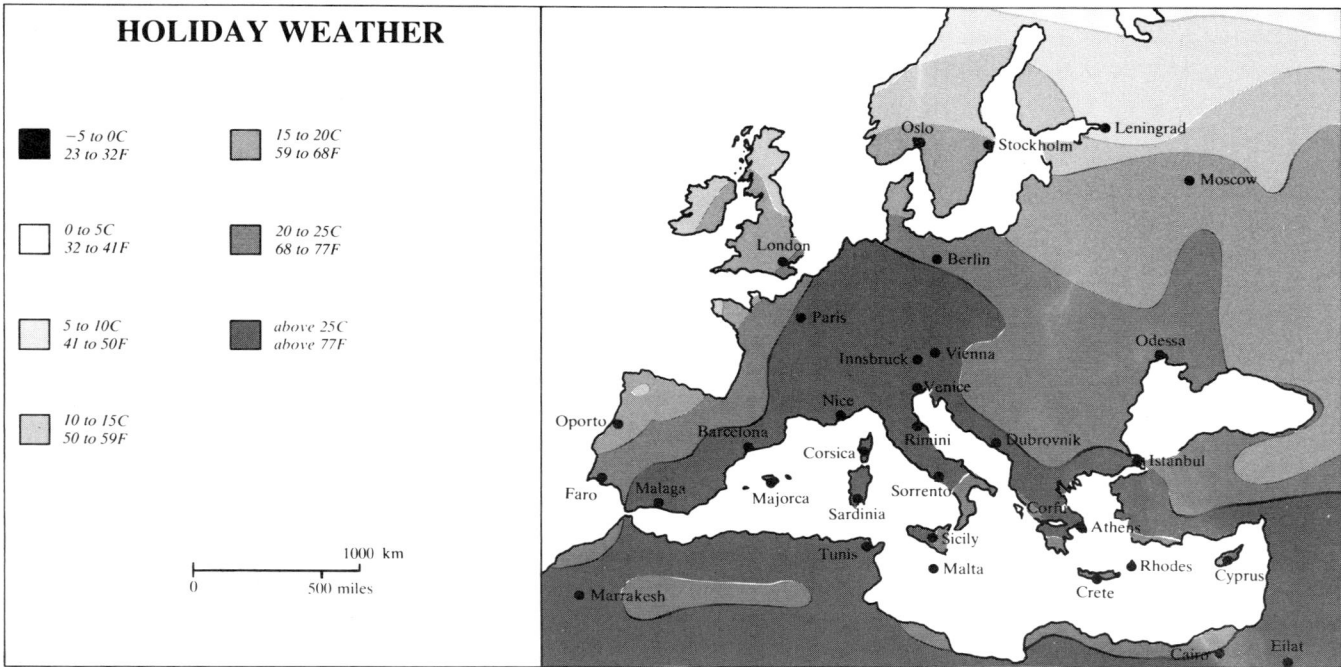

Fig 3.19 ● Holiday weather

The map and data show the temperatures at various places across Europe. What was the temperature in each of these places?

- Dublin
- Berlin
- Stockholm
- Moscow
- Lisbon
- Belgrade
- Vienna
- Helsinki

Spot Check

Can you match each brief clue which follows with a city named on pages 60 or 61?

1. Famous for fashion
2. Most northerly
3. Among the mountains
4. Capital of the smallest country listed
5. Historically the home of democracy
6. Europe's largest port
7. Has neither coastline nor mountains
8. Centre for French, Flemish and Walloon regional assemblies
9. See it from the top of the Tower
10. Only this and Bremerhaven as ports for a big nation
11. Headquarters for many EEC meetings
12. Capital on an island
13. Caesar came from here
14. Hot, with siestas and sombreros
15. Alpine centre
16. On the River Seine
17. Furthest South
18. Vatican City is inside this city
19. Furthest west
20. Over a thousand km from this capital city to the north of its country
21. Largely below sea level
22. Only 8 km from Sweden
23. Near the Iron Curtain

THE WORLD YESTERDAY

EUROPE			
MADRID	sunny 27	TANGIER	fair 29
BELGRADE	fair 26	TUNIS	sunny 28
ATHENS	sunny 26	BIARRITZ	sunny 28
ROME	sunny 25	MARRAKESH	sunny 28
BUDAPEST	cloudy 24	CORFU	sunny 27
VIENNA	fair 23	BORDEAUX	sunny 26
LISBON	sunny 22	SEVILLE	fair 26
ZURICH	cloudy 21	VENICE	sunny 26
BRUSSELS	fair 20	LUGANO	fair 25
WARSAW	cloudy 20	COSTA DEL SOL	fair 24
ISTANBUL	sunny 20	COSTA BLANCA	fair 24
LONDON	fair 20	SICILY	sunny 24
FRANKFURT	rain 19	NICE	sunny 24
PARIS	cloudy 18	DUBROVNIK	sunny 24
MOSCOW	cloudy 18	RHODES	sunny 24
AMSTERDAM	fair 18	INNSBRUCK	cloudy 24
BERLIN	fair 18	TENERIFE	fair 23
COPENHAGEN	fair 17	MINORCA	sunny 23
OSLO	fair 16	GIBRALTAR	sunny 23
HELSINKI	cloudy 14	MALTA	sunny 23
DUBLIN	cloudy 13	CORSICA	sunny 23
STOCKHOLM	cloudy 13	SARDINIA	sunny 23
REYKJAVIK	bright 12	NAPLES	sunny 23
RESORTS:		GRD CANARY	sunny 22
EILAT	sunny 32	COSTA BRAVA	fair 22
FLORENCE	sunny 31	CRETE	sunny 21
MAJORCA	bright 30	MADEIRA	sunny 21
CYPRUS	bright 29	LE TOUQUET	fair 20
		AZORES	cloudy 17

Data by Noble Denton Weather Service

— All temperatures refer to those at local midday —

Quiz

Spot the Errors

What is not correct in the following sentences?

1. It hardly ever rains in Egypt. It is almost always hot and sunny, and so no one can live there.
2. China has about twenty times as many people as Britain. Its famous sights include the Great Wall, the Forbidden City, Red Square and the Temple of Heaven.
3. If you go to the Rockies in North America, you can see the volcanoes in the Grand Canyon.
4. To drive from England to Greece, you must go through France, Germany, Italy and Austria.
5. Since Madrid and Oslo are both about the same distance from London, it cannot be far from Madrid to Oslo.
6. Athens is the capital city of Greece, near the sea, famous for its ancient buildings and only one hour's flight from London.
7. Japan is not very far from the Philippines, China, New Zealand and eastern USSR.
8. It is many times further to go from eastern Australia to South America going past New Zealand than to go from western Australia to South America going past Africa.

Clever Clogs

1. Approximately how many different countries are there in the world? Why is it difficult to be absolutely exact? What makes a country count as an independent state?
2. Which large countries are members of the Commonwealth?
3. How many cities in the world have a larger population than London?
4. What oceans must you cross as you go from continent to continent around the world?
5. With which country in western Europe does the UK trade most?
6. Which country in western Europe has the biggest land area?
7. Which countries in western Europe usually have the warmest temperatures?
8. What three very large nations in the world are federal states?

Picture Search

● This city was the national capital until 1960. Where was the new capital built?

● What big engineering project has made a great difference to this country?

● How large is the area covered by this desert?

Pick the Word

1. Which one of these regions is furthest away from the other three in the list? The Far East / Australia and New Zealand / The Indian sub-continent / North America

2. Which one of these countries is an important oil producer? Israel / Nigeria / East Germany / Cyprus

3. Which of these rivers carries the most water all the year round? Nile / Amazon / Congo / Thames

4. Which is the odd one out in this list, and what is different about it? Tropics / Equator / South Pole / Arctic Circle

5. Which one of these is the most westerly? Austria / Portugal / Yugoslavia / West Germany

6. The area that includes Israel and Lebanon is called the Far East / Middle East / Africa / Asia

7. Which of these countries has much the largest population? New Zealand / South Africa / Bangladesh / Australia

8. Which of these capital cities cannot be reached by sea? Copenhagen / Reykjavik / Amsterdam / Madrid

Crossword

Make a tidy copy of this grid and then solve the clues. Many of the main words have been used on pages 50 to 61 of this book.

Clues across:
1. Not, from, (2).
2. The large country between Portugal and France (5).
5. The water that goes all round the British Isles (3).
6. An island and headland that Britain has had on lease from China for ninety-nine years (4,4).
10. Very large city on the east coast of South America (3).
11. Wight and Man both count as this (4).
12. Last two letters of the word 'tropical' (2).
13. Tomorrow will be, today is, and yesterday ___ (3).
15. Belonging to it (3).
17. Therefore, as a result (2).
19. The continent that includes China and eastern USSR (4).
20. Capital city of China (6).

Clues down:
1. Capital city of Iran (6).
2. First three letters of the word 'segment' (3).
3. A large country lying north-west of India (8).
4. One of the largest African states, and a large oil producer (7).
5. When this melts in the Ethiopian Highlands, the water runs down the Nile to Egypt (4).
7. A valuable product from the North Sea, Nigeria, North Africa and the Middle East (3).
8. The way we normally write 'Ordnance Survey' maps (2).
9. Short for the Netherlands on a car's registration (2).
14. Vowels in the word 'Spain' (2).
16. Consonants in the word 'sausage' (3).
17. Short for 'south-east' (2).
18. All right, good enough (2).

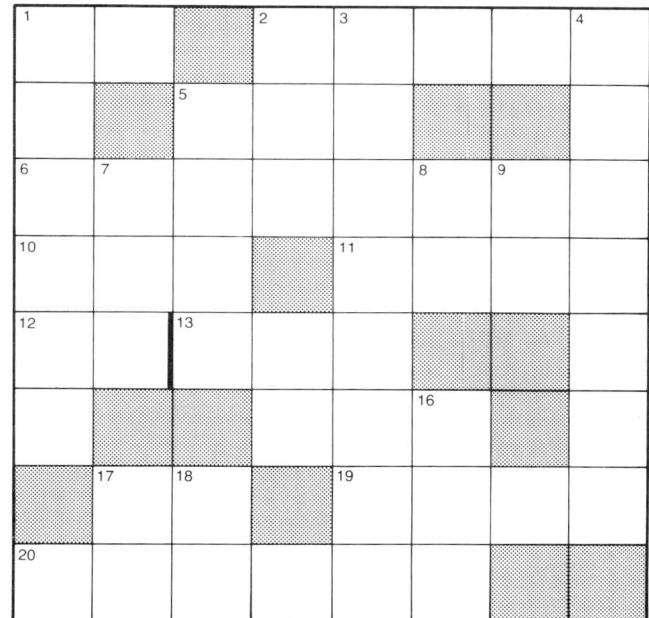

Inland Waterways

Millionaire Wants a Change

An American pop star wants to come to Britain to spend a quiet week unnoticed in the countryside, to escape for a while from the glare of publicity. A cruise on a canal boat seems to be the ideal answer. You work in an agency booking cruises. The pop star's list of questions has been given to you. Prepare your answers to his questions, using the information that follows and reference books.

Britain's Canals

Two hundred and fifty years ago, the only way to travel or to move goods through Britain was on foot or with a horse. The roads were full of potholes, ruts and mud. Even heavy loads like corn, building stone, iron or coal had to be moved this way. Carrying a load from one big town to another could take several days.

Long before the railways were invented, engineers began building canals. One horse could pull a much greater load walking along a towpath pulling a barge than walking on a road pulling a cart. The horse could walk faster on the towpath because it was not full of ruts made by the wheels of carts.

The only real problem in building canals between all Britain's big cities was that they had to go over or round hills. The engineers tried to choose flat routes as much as possible. When they needed to go up or down a hill they built locks (see page 87 for a photograph). The biggest ladder of locks was on the Kennet and Avon Canal at Devizes in Wiltshire, where the boats had to go up or down through sixteen locks in a row.

The canals were the main system of freight transport for over fifty years, until the railways were developed. Some canals have been neglected or removed, but there is still a national system with nearly 2000 miles of canals and rivers through which you can take a boat. The canals are used a great deal for holidays and 340 miles are still used for freight transport. The map shows the total extent of navigable inland water in Great Britain. The canals were built to link with the rivers, and lead into and out of rivers at some points.

Fig 4.1 ● Canal boat on the Staffordshire and Worcestershire canal

Pop Star's Questions

You will need an atlas of Britain showing details of hills, rivers and the big cities to supply the extra information needed to answer these questions:

1. Where do canals join the rivers Severn, Thames and Trent?
2. The city of Birmingham is in a hilly district and mostly more than 150 metres above sea level. It seems a difficult place to build a canal. Why were so many canals built there? Is it worth travelling on them?
3. There are some other ranges of hills between London and Birmingham. Do the canals go over or past them?
4. Which other rivers besides those named in question 1 are linked with the canal system?
5. Is it true there is a canal from the Manchester area to the district around Leeds and Bradford? Why did the engineers build a canal in such a difficult area?
6. Why are there not many more canals in East Anglia and Lincolnshire, since the land there is so flat?
7. Why aren't there any canals around Dover?
8. If I take a canal boat for a holiday cruise, will it have an engine or be pulled by a horse? I have been told I am not allowed to go faster than 4 mph. Is that true? If so, why are speed controls necessary on canals?

9. If I took a canal boat for a week's holiday cruise and decided to travel for about eight hours a day (including going through some locks), how far would I be likely to travel in one day? How far would this be in a week?

10. I am thinking of starting from the following cities. Would I be able to complete a canal tour that started one way out, and come back by another way, within one week? (a) Oxford (b) London (c) Leeds

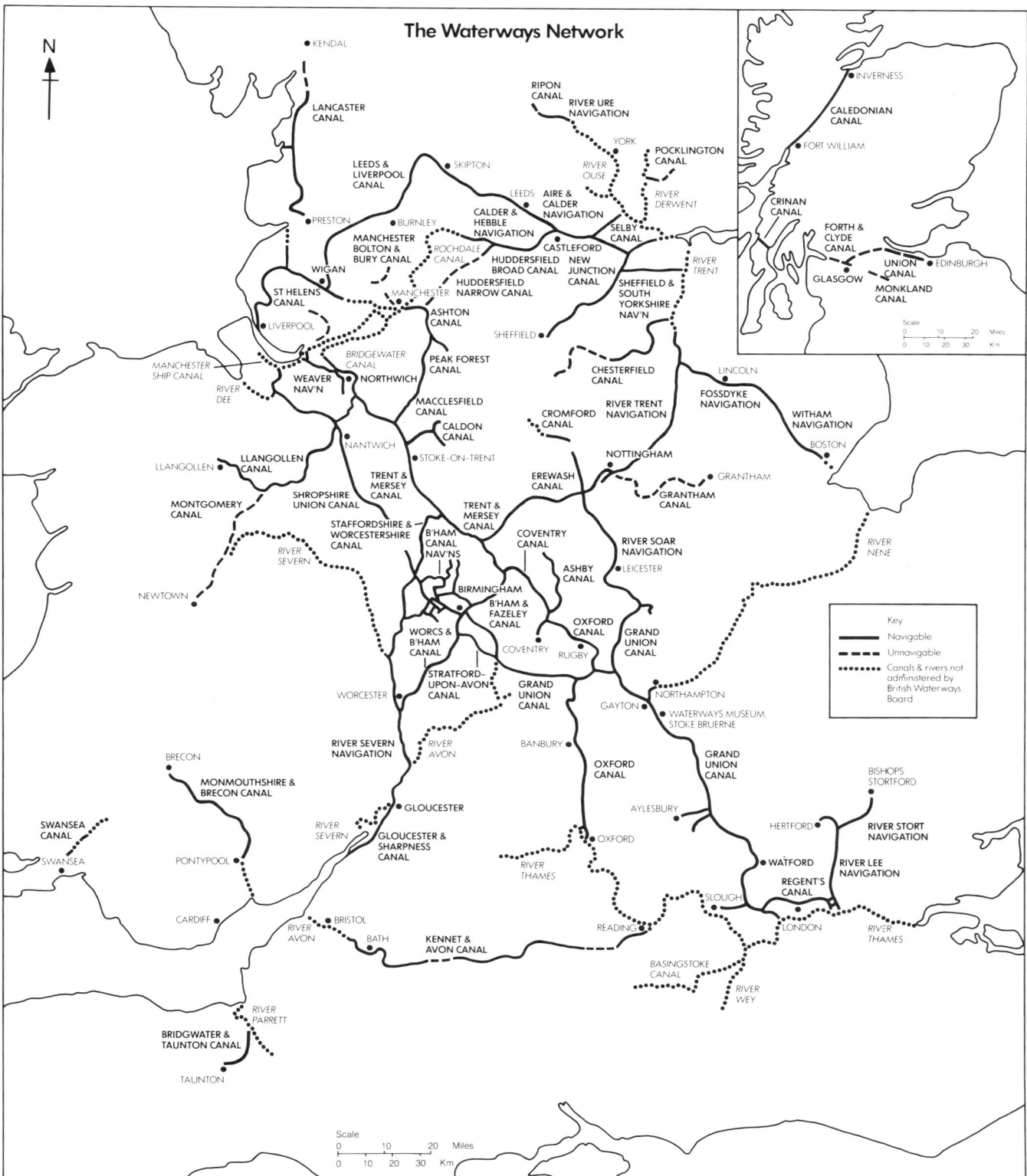

Fig 4.2 ● Waterways network of Britain

Celtic Tour

What is needed to fill the gaps in this account of an adventurous tour of western Britain – the Celtic homelands?

You may need an atlas or set of maps to help you.

Angus and Fiona set off from Inverness and drove south, leaving the (1)____ hills on their left with the clouds hanging over them. They went past Dundee and Perth and came in lovely sunshine to (2)____, the capital city of Scotland. Here they crossed the Firth of (3)____ and then drove along the (4)____ motorway to Glasgow. By now the rain was beginning. They went on and caught the ferry from Stranraer to Larne in (5)____.

From Larne, they went to (6)____, the capital of the province, and then to Ireland's largest lake, Lough (7)____. Driving south again, they reached the Mourne Mountains. After that they crossed the border into (8)____.

Continuing south, they came to Dublin, the (9)____ of Eire, and took a ferry from near here to Holyhead on the island of (10)____. Crossing the Menai Straits, they admired the highest mountains in Wales, in the (11)____ National Park.

Driving generally in a (12)____ direction along the west coast of Wales past Aberystwyth, they came at last to the Pembrokeshire National Park in the county of (13)____. The sandstone cliffs were warm in the sunlight.

From Pembroke, they set off east to join the (14)____ motorway near Swansea, passed (15)____, the capital of Wales and crossed the River (16)____ on a high suspension bridge.

Finally they turned north on the (17)____ motorway, and joined the M6 after Birmingham. This took them all the way north as far as (18)____ in the county of (19)____. Soon after that they were able to cross the border into Scotland again. Later they made a quick stop to look for the monster in (20)____ before returning to Inverness.

Fig 4.3 ● View of Snowdon from Llyn Llydaw lake

Crack the Code

Fig 4.4 ● Bradford

SRHSTB

This code gives the first letters of six towns in these counties: Yorkshire (North, South and West), Cleveland, Durham, Tyne and Wear and Northumberland. Three teams had to decide (1) what the towns were (2) what they had in common and (3) what would be the shortest route to visit them all. Team B won because the towns were in order along a direct route and they all had something in common. Using a map to check their route, explain why their answers were better than the others.

Team A chose: Selby, Rotherham, Halifax, Sheffield, Tynemouth, Bradford

Team B chose: Scarborough, Redcar, Hartlepool, Sunderland, Tynemouth, Berwick

Team C chose: Stockton, Richmond, Halifax, South Shields, Thornaby, Bishop Auckland.

Snowdonia

Use a map of Snowdonia to help you. Plan good routes for each of these groups and say what towns they would pass through:

Group One. Students on a day visit from Bristol want to go to the top of Snowdon, see the Menai Straits and study the effect of water power by visiting the Trefriw Woollen Mills (10) and the Ffestiniog Hydro Electric Scheme (5).

Group Two. Other students travelling from Leeds to Ireland want to stop on the way to visit the Porthmadog Pottery (2), to Gloddfa Ganol Slate Mine (6), the Snowdon Mountain Railway (7) and the Llanberis Lake Railway (8) before taking the Sealink ferry (11).

Group Three. A family travelling by road from Stoke on Trent want to go to the Butlins Day Visitor Centre (4), Gloddfa Ganol Slate Mines (6), Penrhyn Castle (9) and Criccieth Castle (3). They will spend one night in the area.

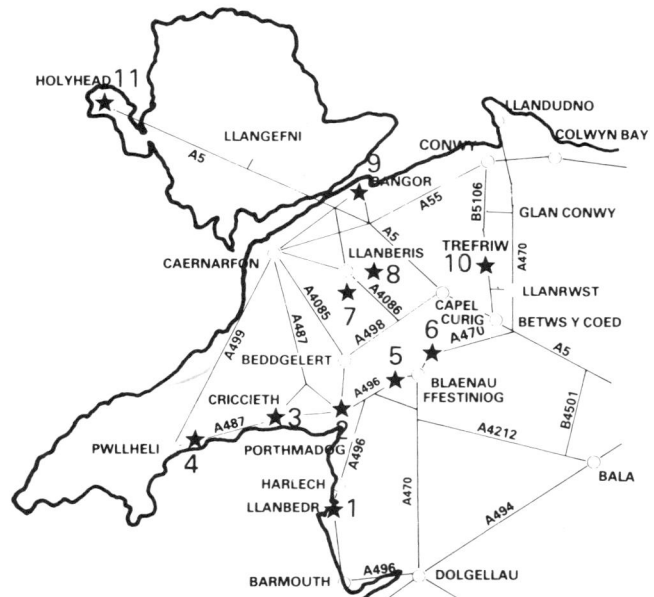

Fig 4.5 ● Map of Snowdonia

Pictures and Maps

The photographs on page 69 show nine geographical features. The diagram below shows symbols used on maps to show these features. Here are brief descriptions. Try to link up each photograph with the right symbol and the right description.

Photographs: R S T U V W X Y Z
Symbols: ① ② ③ ④ ⑤ ⑥ ⑦ ⑧ ⑨

Descriptions:

Mountain pass A place where the hills are not so high, and it is possible to travel through without going over the very steep slopes.

Moorland Wild open country with low tough plants like heather, bilberry, rush, and coarse grass. No trees.

Ford A place where the road passes through a stream, but the ground is firm and the water is not too deep.

Mountain farm An isolated farm on a mountain side. It is unlikely to have many fields, but the animals will wander far over the unfenced hillsides.

Marsh A flat area where there is little drainage and the water lies on or near the surface almost all the year round. The ground may be very soft and anyone walking on it may be trapped by sinking into the mud. Reeds, bog cotton and other marsh plants thrive.

Weir A place where a barrier has been built across a river to hold back some of the water above, usually to keep it at a particular depth. The rest of the water pours over the weir in a waterfall. If it is intended that boats should pass by, there will be a lock beside the weir.

Bay An area of coastline, usually calm, where the sea is sheltered by land on both sides. Bays often have sandy beaches because the storms do not hit them so hard.

Estuary The place where a river joins the sea. The water may be tidal, which means that the level of the river will rise and fall with the tide. There may be mud flats and areas of deep water may be marked with buoys or beacons.

Bridle path A path wide enough for a horse to pull a cart but where motor vehicles are not allowed. Most bridle paths are not surfaced.

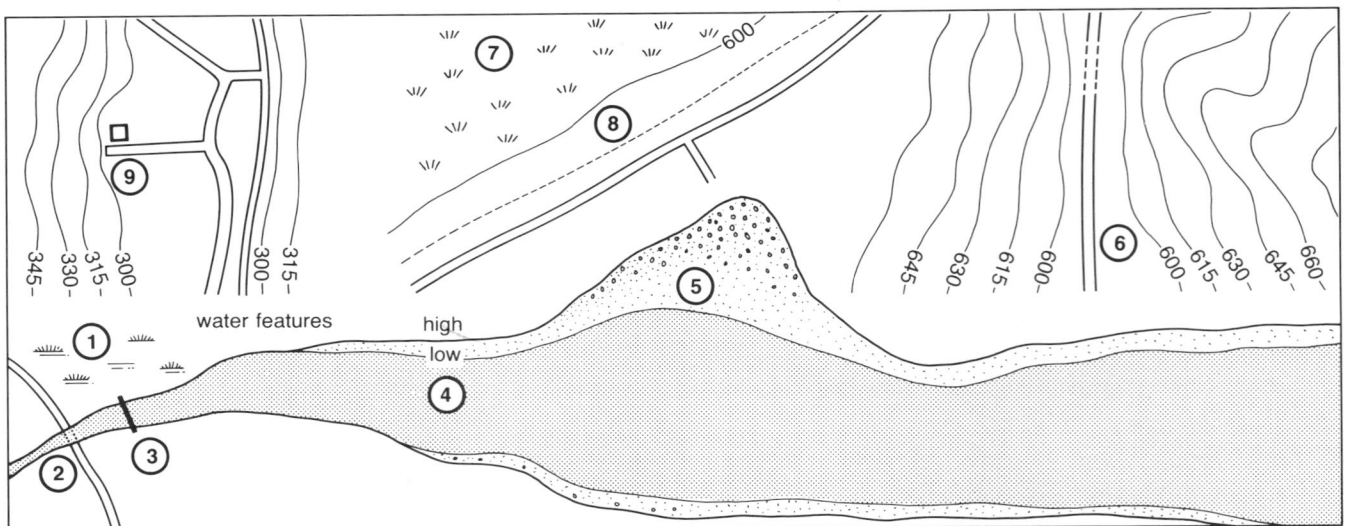

Fig 4.6 ● Symbols used to represent certain features on maps. Match the numbers to the descriptions above and the photographs on page 69

R

V

W

S

T

X

U

Y

Z

Weather on the Map

Britain is not one of the world's large countries, but the weather may vary greatly from one part to another. Britain's weather also changes far more often than it does in many other parts of the world. It is not surprising that British people are always talking about the weather. There is often more to say about Britain's weather, compared to some other countries.

Suppose your birthday started off with telephone calls from no less than six relations. As usual, they talked about what they thought the weather would be like today. What was it they all said? Use the map to find out what comments each was likely to make.

- Your aunt in Norwich said:
- Your cousin in Aberystwyth said:

- Your friend in Plymouth said:
- Your great uncle in Glasgow said:
- Your other cousin in Birmingham said:
- Your grandparents in Edinburgh said:

1. What weather were you expecting where you live?

2. What weather would someone travelling from Cardiff to Hull expect on the journey?

3. What weather would someone travelling from Dublin, through Holyhead to Dover, expect on the journey?

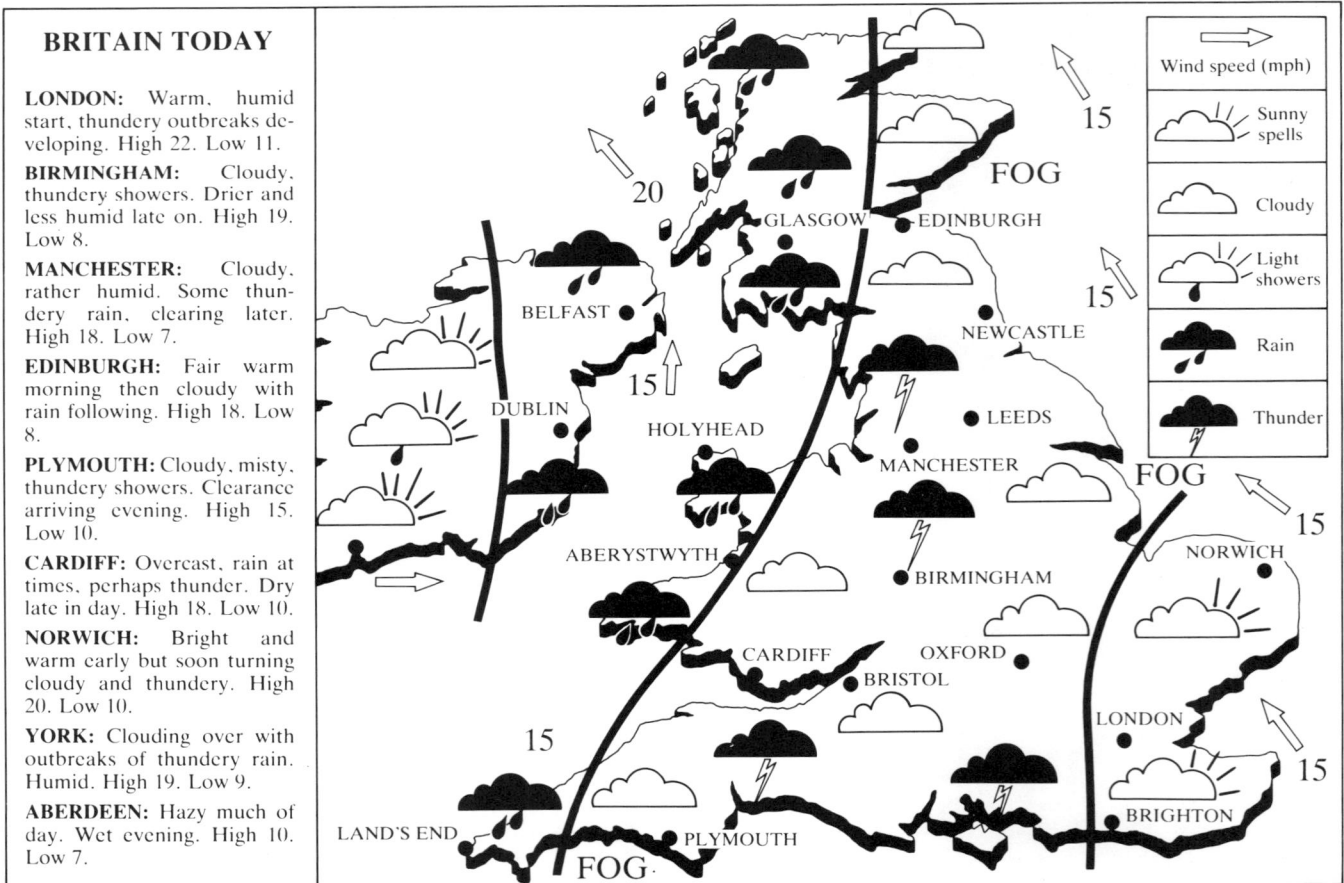

BRITAIN TODAY

LONDON: Warm, humid start, thundery outbreaks developing. High 22. Low 11.

BIRMINGHAM: Cloudy, thundery showers. Drier and less humid late on. High 19. Low 8.

MANCHESTER: Cloudy, rather humid. Some thundery rain, clearing later. High 18. Low 7.

EDINBURGH: Fair warm morning then cloudy with rain following. High 18. Low 8.

PLYMOUTH: Cloudy, misty, thundery showers. Clearance arriving evening. High 15. Low 10.

CARDIFF: Overcast, rain at times, perhaps thunder. Dry late in day. High 18. Low 10.

NORWICH: Bright and warm early but soon turning cloudy and thundery. High 20. Low 10.

YORK: Clouding over with outbreaks of thundery rain. Humid. High 19. Low 9.

ABERDEEN: Hazy much of day. Wet evening. High 10. Low 7.

Fig 4.7 ● Weather map of Britain

Details of Climate

Figure 4.8 shows what sort of weather you can expect month by month in three places in the Far East. The weather details for Wakefield in the UK are included for comparison.

Various people want advice about when they should visit one or other of these three places in the Far East. What would you recommend?

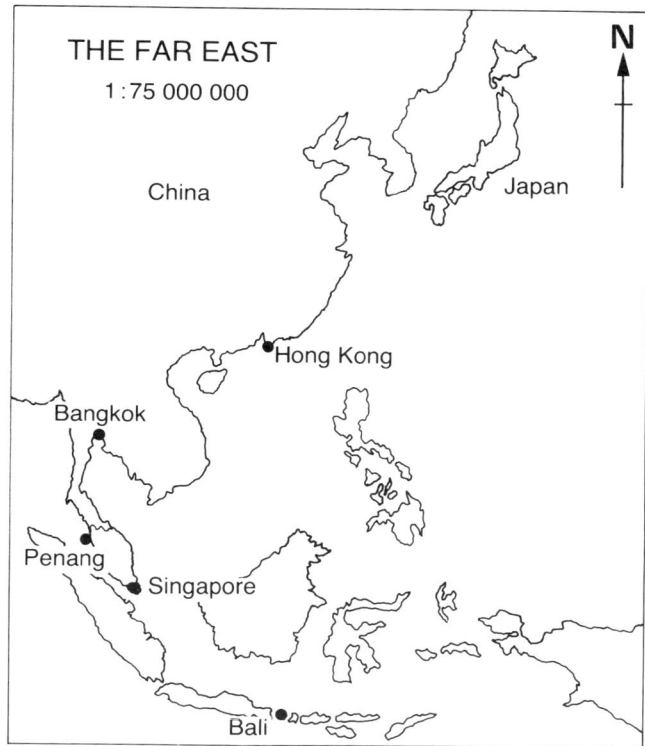

CLIMATE Singapore

Nov	Dec	Jan	Feb	Mar	Apr	May	Jun	Jul	Aug	Sep	Oct
86	86	86	87	88	88	87	87	87	86	87	87

Average daily maximum temperature °F

5	4	5	6	6	6	6	6	6	6	6	5

Average hours of sunshine per day

10	10	10	7	7½	7½	7	7	7	8	7	8

Average monthly rainfall in inches

CLIMATE Penang

Nov	Dec	Jan	Feb	Mar	Apr	May	Jun	Jul	Aug	Sep	Oct
87	87	89	90	90	90	89	89	89	88	88	88

Average daily maximum temperature °F

6	7	7½	8	8	7½	6	6½	6½	6	5½	5

Average hours of sunshine per day

12	6	4	3	6	7	11	8	7	12	16	17

Average monthly rainfall in inches

CLIMATE Bali

Nov	Dec	Jan	Feb	Mar	Apr	May	Jun	Jul	Aug	Sep	Oct
90	87	87	88	87	88	87	86	85	86	86	86

Average daily maximum temperature °F

10	10	8	10	10	10	9	9	9	10	11	10

Average hours of sunshine per day

10	12	15	12	9	6	5	6	5	3	2	5

Average monthly rainfall in inches

CLIMATE Wakefield

Nov	Dec	Jan	Feb	Mar	Apr	May	Jun	Jul	Aug	Sep	Oct
48	44	43	44	48	54	61	66	68	66	64	57

Average daily maximum temperature °F

2	2	1	3	4	4	5	6	7	7	3	3

Average hours of sunshine per day

2	2	3	3	2	2	3	2	3	3	2	3

Average monthly rainfall in inches

Fig 4.8 ● Climate in the Far East and Wakefield

THE FAR EAST
1 : 75 000 000
N

China
Japan
Hong Kong
Bangkok
Penang
Singapore
Bali

- Kausar is determined to go to Singapore, and wants as much sun as possible

- Nathan is a scientist wanting to make a special study of effects of heavy rain on insect life

- Scott does not like it too hot, but does like long hours of sunshine

- Kashif wants to see the Far East, but he does not like much sunshine

- Sorrell wants specially to see Bali, and does not want much rain

- Ailsa likes heat and sun, and hates rain

The Lake District

Janine and Toni, Kristian and Tyron are keen members of the YHA, and they have decided they would like a walking holiday in the Lake District. There are some things that they particularly want to see. They know that if they are going to have time to see these, and to enjoy their day without too much travelling, it is enough to plan to cover only about ten miles a day. They are going to start from Kendal about midday on Monday, and have to get back to Kendal on Friday afternoon. They can have a total of four nights in the hostels in the district.

Use this map to help them plan the best order for seeing the different things on their list. Which hostels should they book into, and for which nights should they book?

The four want to go to the Steamboat Museum on Windermere, and to go round the Wildlife Park at Lowther. They also want to travel on the Ravenglass and Eskdale Steam Railway. They want to climb to the top of Helvellyn, Langdale Pikes and Scafell Pikes and they want to enjoy the views over Ullswater and Buttermere. They plan to visit a friend in Haverthwaite and they want to take photographs at Levens Hall.

Fig 4.9 ● Boat on Lake Windermere

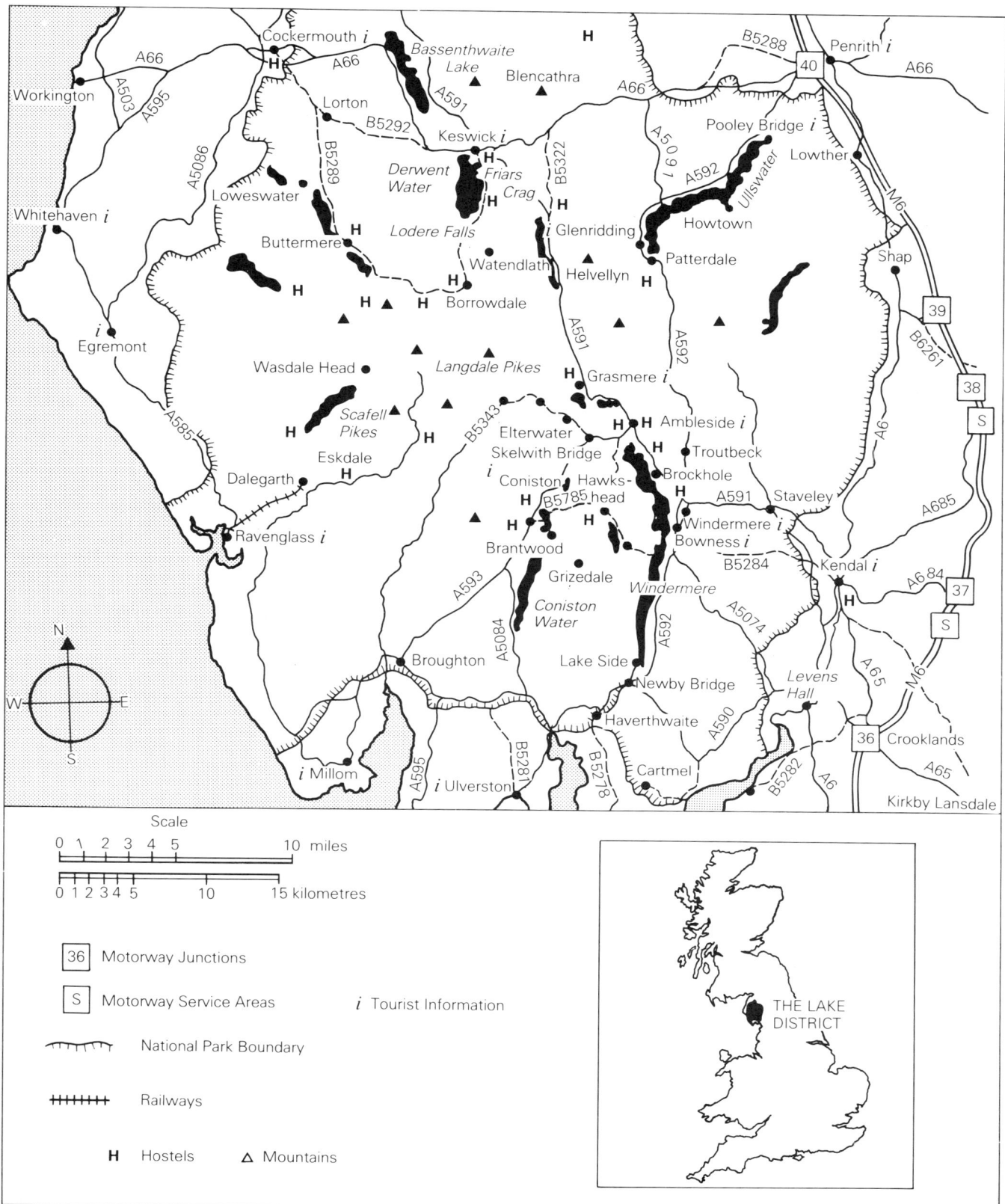

Workington
A66
A503
A595
Cockermouth *i*
A66
Bassenthwaite Lake
Blencathra
H
B5288
Penrith *i*
40
A66
Lorton
B5292
A591
Keswick *i*
A5091
Pooley Bridge *i*
Lowther
M6
Whitehaven *i*
B5289
Loweswater
Derwent Water
Friars Crag
B5322
Glenridding *i*
Ullswater
Howtown
A592
Shap
H
Buttermere
Lodere Falls
Watendlath
Helvellyn
Patterdale
39
Egremont *i*
Borrowdale
A591
A592
B6261
Wasdale Head
Langdale Pikes
Grasmere *i*
38
S
A5086
Scafell Pikes
B5343
Ambleside *i*
A6
Eskdale
Elterwater
Skelwith Bridge
Troutbeck
Brockhole
A585
Dalegarth
Coniston
Hawkshead
B5785
Staveley
A591
A685
Ravenglass *i*
Brantwood
Windermere *i*
Bowness *i*
Kendal *i*
A6 84
37
Grizedale
B5284
S
A593
Coniston Water
Windermere
A5074
A65
M6
Broughton
A5084
A592
Lake Side
Newby Bridge
Levens Hall
36
Crooklands
A590
Haverthwaite
A65
Millom *i*
A595
B5281
Ulverston *i*
B5278
Cartmel
B5282
A6
Kirkby Lansdale

N
W — E
S

Scale

0 1 2 3 4 5   10 miles

0 1 2 3 4 5   10   15 kilometres

| 36 | Motorway Junctions |
| S | Motorway Service Areas | *i* Tourist Information |

National Park Boundary

Railways

H Hostels   △ Mountains

THE LAKE DISTRICT

Fig 4.10 • Map of the Lake District

73

The Netherlands (Holland) and Beyond

What are the best routes you can work out for these three people? They will all travel by road, and cross by ferry from Harwich to the Hook of Holland.

1. Melinda lives in Guildford and she wants to visit the following places. She can be away for only three nights and she wants to stay each night at one of the places shown on the map. What places would you advise her to visit each day, and where would you advise her to stop each night? Give full road directions for her route. The places she wants to visit are:

- Antwerp
- Bremen
- Dortmund
- Eindhoven
- Groningen
- Koln
- Utrecht

2. Stefan lives in Portsmouth. He needs to visit factories in the following towns and will travel to and from the Continent via the Hook of Holland. Give details of the best routes he can take across England and on the Continent. How far from the Hook of Holland is the furthest of the towns he will visit?

- Utrecht
- Dusseldorf
- Rotterdam
- Hamburg
- Koblenz

3. Georgina lives in Shrewsbury. What is her best route to Harwich? She has booked rooms in a hotel in Hook of Holland for herself and her elderly mother and wants to make day visits from there to as many other European countries as possible. Her mother does not want to travel more than 250 miles altogether in any one day and wants to be able to return to the same hotel each evening. They are thinking of visiting the following places. Which ones can they visit?

- Brussels
- Dortmund
- Trier
- Amsterdam
- Eindhoven
- Bonn
- Kiel

How many countries will they have visited in the end?

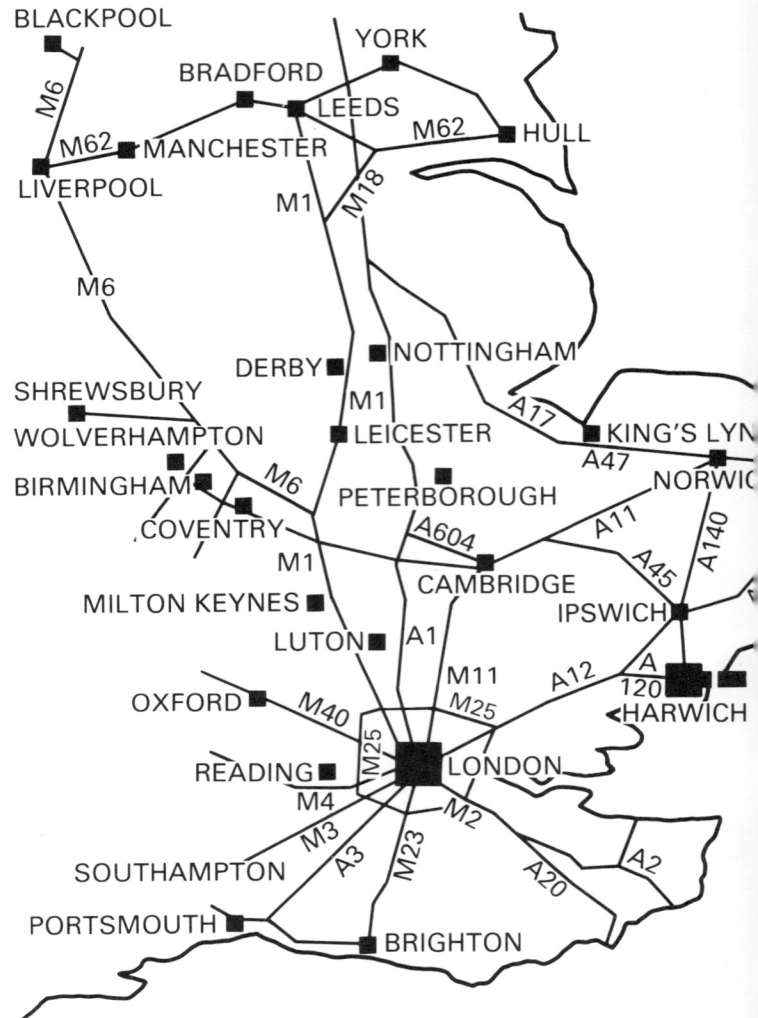

Fig 4.11 ● Main road routes in Britain and Europe

MILEAGE GUIDE from the Hook

1 mile – 1.6 km 1 km – 0.6 miles	TOTAL ROAD MILEAGE	MOTORWAY MILEAGE INCLUDED
HOLLAND		
AMSTERDAM	46	35
ROTTERDAM	15	10
HAGUE	10	5
UTRECHT	53	46
EINDHOVEN	78	70
GERMANY		
COLOGNE (KOLN)	187	180
DUSSELDORF	172	165
FRANKFURT	292	285
HAMBURG	303	212
HEIDELBERG	348	340

HARPENDEN
A1 A414 A10 A414 HARLOW
M11
M1 HODDESDON
HATFIELD M25 EPPING
M40 RICKMANSWORTH WALTHAM CROSS A414
A12 TO HARWICH
NEW CHELMSFORD BYPASS
M4 M25 A127
LONDON A13
WINDSOR STAINES GRAYS
A30 DARTFORD A2
M3 M25 M25
GUILDFORD LEATHERHEAD SEVENOAKS A25
DORKING M25 M26
A3 REIGATE A25 A25 TONBRIDGE
A3 A25 A24 M23 A22 A21

KIEL
TO DENMARK, SWEDEN & FINLAND
HAMBURG
E35 E3
BREMEN
E72
E3
GRONINGEN A7
E35
A28
E35 A30
AMSTERDAM
THE HAGUE E10 E9 E35
E9 UTRECHT
HOOK OF HOLLAND
ROTTERDAM E8 E36
ARNHEM
A16 A15 E73
A37 DORTMUND
BREDA A3 A44
EINDHOVEN A1
E3 A45
ANTWERP E10 E3
E10 DUSSELDORF
KOLN
BRUSSELS BONN
A3 E4
KOBLENZ E4
A48 FRANKFURT
BOPPARD
TRIER
TO SWITZERLAND AUSTRIA & ITALY

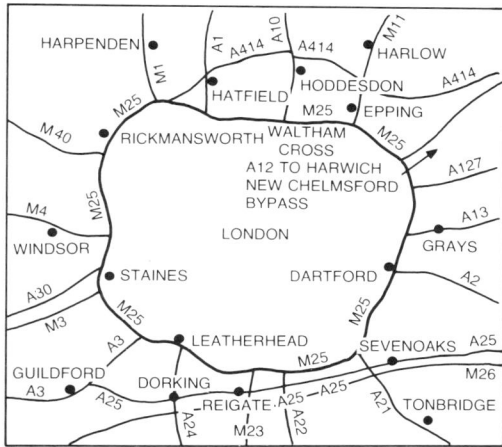

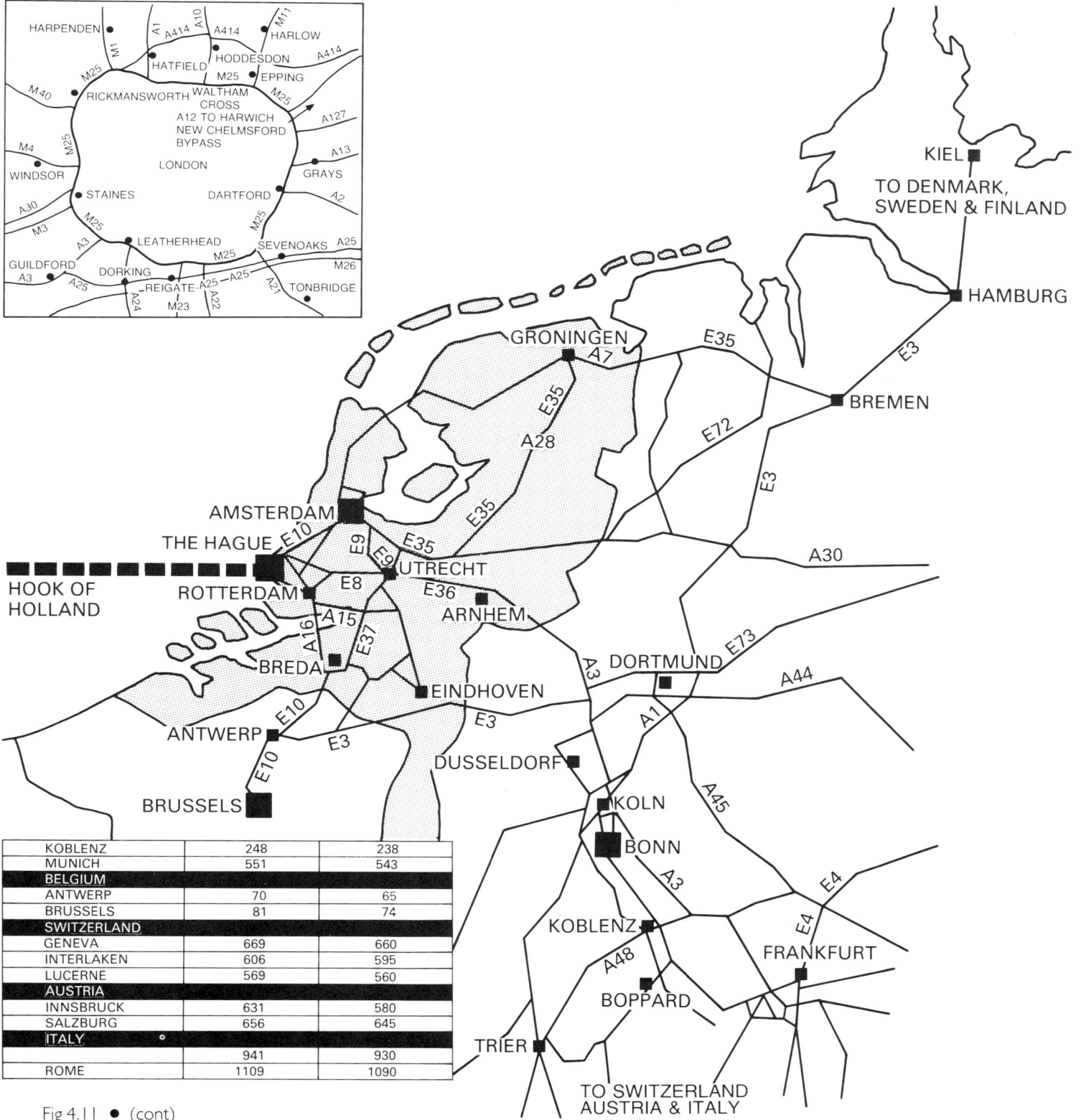

KOBLENZ	248	238
MUNICH	551	543
BELGIUM		
ANTWERP	70	65
BRUSSELS	81	74
SWITZERLAND		
GENEVA	669	660
INTERLAKEN	606	595
LUCERNE	569	560
AUSTRIA		
INNSBRUCK	631	580
SALZBURG	656	645
ITALY		
	941	930
ROME	1109	1090

Fig 4.11 ● (cont)

Problem Solving: Europe

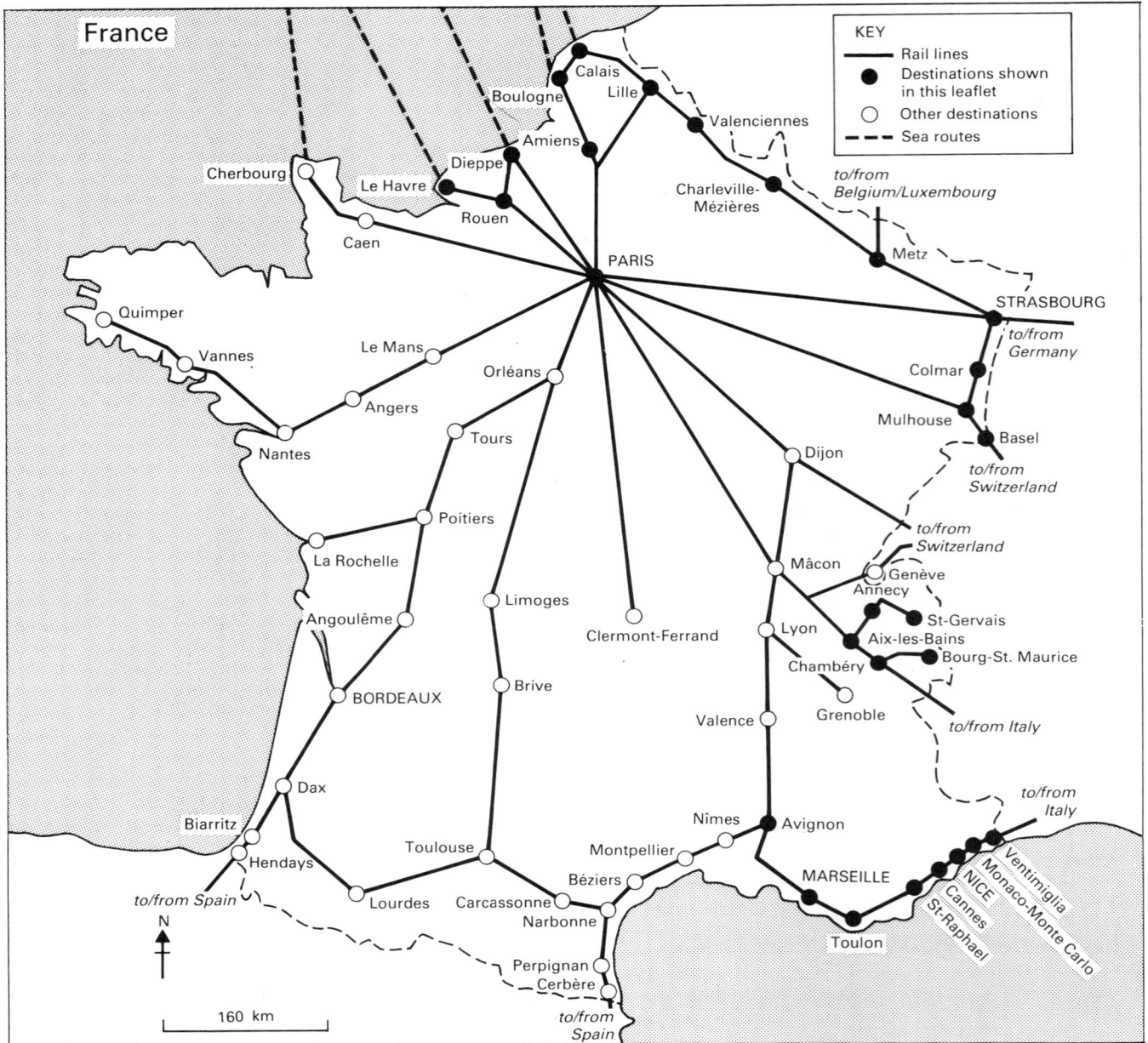

Fig 4.12 ● Main rail routes in France

Cheap Week

You have bought a cheap ticket that allows you to go to Paris and back, with unlimited rail travel in France during one week in March. Plan a week that will allow you to do all these things:

(a) go skiing (b) see the Mediterranean (c) spend a day sightseeing in Bordeaux (d) visit a friend who is working in Strasbourg (e) go climbing in the Pyrenees (f) go up the Eiffel Tower.

You should aim not to travel more than 480 km on any one day.

N

Fig 4.13 ● Some European road routes

Entry Test

You think you might like to work for a Travel Agent. When you apply for the job, the agent sets you this test:

A customer in Sheffield has asked us to plan a route for a lorry that has to drop heavy loads in the following cities during January. We are to advise of possible delays. Make the plan, and explain why you chose the routes concerned. Comment on possible delays on the route chosen, weather conditions and frontier controls. The loads are to be dropped at Bern, Hamburg, Munchen, Nice, Rome, Rotterdam.

Weather report: Snow has fallen overnight across southern France and northern Italy, but roads in Germany are still clear. The passes over the Alps are all open, but some delays are likely.

Top-Sales Ltd want to deliver leaflets for distribution to every passenger on the flights shown in Fig. 4.14. They have four vans that will deliver the leaflets directly to the airports concerned in England, Wales and Scotland. Leaflets for Northern Ireland will be sent by normal postal services.

The leaflets have been printed in two languages – English, and the language that is spoken where the plane will land. Top-Sales Ltd plan to use 300 leaflets on each flight, and to keep up their sales campaign for ten days. That means that they will need 3000 copies of each leaflet for each flight shown here.

Work out how many of each sort of leaflet should be put in each van. The vans are allocated airports as follows:

Sherpa van: Airports in the London area
Bedford van: Airports in Scotland, and England north of Yorkshire
Ford van: Airports in East Anglia, the Midlands including Yorkshire
Renault van: Airports in Wales and the west of England.

Fig 4.14 ● This chart shows which European airports can be reached on direct flights by one holiday operator from UK airports. For the exercise assume that there is one flight a day from each airport

		GATWICK	HEATHROW	LUTON	STANSTED	NORWICH	BRISTOL	CARDIFF	BIRMINGHAM	EAST MIDLANDS	MANCHESTER	LEEDS/BRADFORD	NEWCASTLE	GLASGOW	EDINBURGH	BELFAST	ABERDEEN
SPANISH MAINLAND & GIBRALTAR (SPANISH)	Costa Brava	✈	✈						✈	✈	✈	✈	✈	✈	✈		✈
	Costa Dorada	✈	✈						✈	✈	✈	✈	✈	✈	✈		
	Costa Blanca	✈	✈						✈	✈	✈	✈	✈	✈			
	Costa Del Sol	✈	✈	✈					✈	✈	✈	✈		✈	✈		
	Gibraltar – Twin Centre	✈	✈	✈					✈	✈	✈			✈	✈		
CANARY ISLANDS (SPANISH)	Tenerife	✈	✈						✈	✈	✈	✈		✈			
	Gran Canaria	✈	✈								✈	✈		✈	✈		
BALEARIC ISLANDS (SPANISH)	Minorca	✈	✈								✈	✈	✈				
	Ibiza	✈	✈						✈	✈	✈	✈	✈	✈	✈		
	Formentera	✈	✈							✈	✈	✈					
	Majorca	✈	✈	✈	✈	✈	✈		✈	✈	✈	✈	✈	✈	✈	✈	✈
PORTUGAL & MADEIRA (PORTUGUESE)	Estoni Coast	✈	✈								✈						
	Algarve	✈	✈					✈		✈		✈		✈	✈		
	Madeira	✈															
NORTH AFRICA (ARABIC)	Tunisia	✈	✈						✈	✈	✈						
	Morocco – Tangier	✈	✈								✈						
	Morocco – Agadir	✈															
ITALY & SARDINIA (ITALIAN)	Adriatic Riviera	✈	✈							✈	✈		✈	✈			
	Venetian Riviera	✈	✈								✈			✈		✈	
	Ligurian Riviera		✈														
	Neapolitan Riviera	✈	✈							✈		✈		✈			✈
	Sardinia	✈									✈						
MALTA (MALTESE)	Malta	✈	✈	✈						✈		✈		✈	✈		
GIBRALTAR (ENGLISH)	Gibraltar	✈															
GREEK MAINLAND (GREEK)	Greek Mainland	✈	✈							✈	✈			✈			
	Halkidiki	✈	✈								✈						
GREEK ISLANDS (GREEK)	Aegean Islands	✈	✈							✈	✈			✈			
	Tinos	✈															
	Corfu	✈	✈					✈	✈	✈	✈	✈	✈	✈	✈		✈
	Kefalonia	✈									✈						
	Zakynthos	✈									✈						
	Kos	✈	✈								✈			✈			
	Rhodes	✈	✈						✈	✈	✈	✈		✈			
	Crete	✈	✈						✈	✈	✈	✈		✈	✈		

Super Competition

The Grand Slam competition you have won allows you and a friend to fly around the world. You may choose your own route so long as you keep going in the same general direction all the time. You can use only the air routes shown on this map, starting from London.

You are allowed six stop-overs at any six of the points shown on this map. At each stop-over your prize will include travel to any place within 250 miles of the airport. You cannot afford to pay any other costs, so you have to keep within these limits.

You and your friend have made a list of the wonderful places you may visit, but you realise you cannot visit them all and some of them are not near enough to the airports on the map. Find out as much as you can about these places, and then choose your six stop-overs to allow you an exciting and varied tour. Why did you choose these places?

Fig 4.15 ● Some air routes of the world

The list you have made is:
- **Egypt and the Nile** – Cairo, Luxor, Aswan Dam
- **Indian Ocean** – Seychelles, Mauritius
- **Hong Kong** – Kowloon and small islands
- **Singapore** – on the Equator, the world's fourth largest port
- **Pacific Ocean islands** – Hawaii, Fiji, Tahiti – coral reefs, volcanic peaks and sunshine
- **China** – Peking, the Great Wall and Chinese culture
- **Japan** – Tokyo, Mount Fuji, monorail trains
- **Alaska and Canada** – spectacular wild countryside
- **USA** – Florida, Disney World, Kennedy Space Centre, Hollywood
- **Brazil** – Iguacu Falls, Amazon jungle, and Rio de Janeiro
- **New Zealand** – hot springs and geysers, beautiful lakes
- **India** – Himalayas, Taj Mahal
- **Central America** – Mexico, the Panama Canal, the Caribbean, Haiti, Bahamas

Simulation Game

In a new sort of treasure hunt, three teams are competing for a big prize. They have to find and bring back chunks of special rock placed at three points in Iceland: Husafell (number 5 on this map), Askja (number 46) and Kerlingsfjoll (number 78). The three teams each have their own ideas on which way will be fastest.

What you have to do is to follow their routes, and work out who was the winner. You might do best to take a team each, work out your time, and see if you can claim the prize. Your note of how long you took can be challenged by either of the other teams.

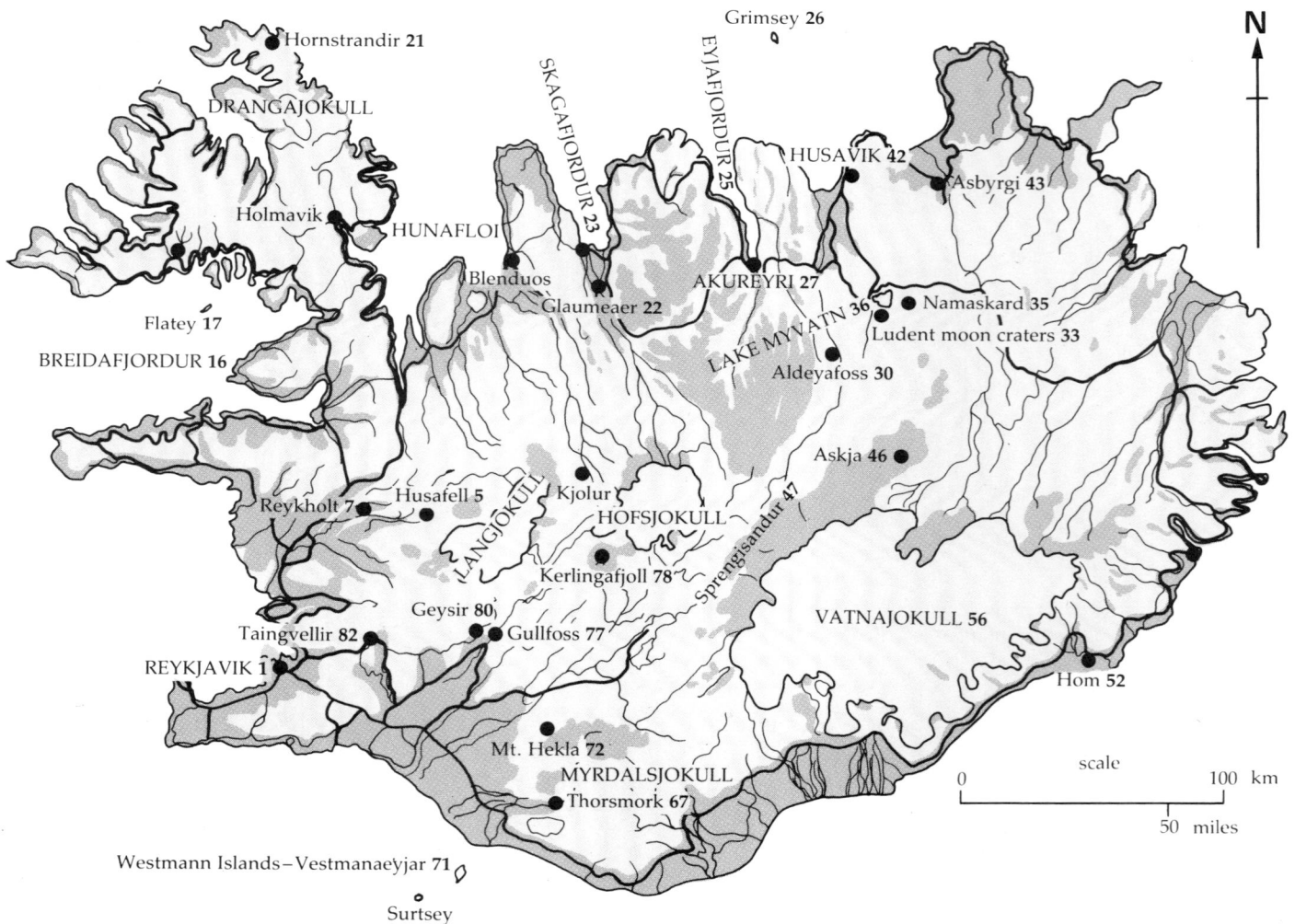

Fig 4.16 ● Map of Iceland

These are the rules:

The teams set off at the same time from Reykjavik (number 1 on the map) at 0600 on 21 June. At this time of year the weather will be fairly warm, and the teams will not be delayed by snow at any point on their route. They may either cycle along the roads, which are made of gravel, sand and lava dust, or they may hike across the open moorland.

You can assume they will cover 150 km a day by cycling, but note, the roads are not straight! When hiking across moorland they will normally cover 40 km a day. But if it is high land (grey shading), they will only manage 20 km a day. Every time they must cross a stream, take off 10 km from their day total. To work out how far they go each day, use pieces of coloured cotton measured to one day's journey for each method of travelling.

The three ice fields, Langjokull, Hofsjokull and Vatnajokull, are all out of bounds. The teams must go round them, not across them. Teams will not have time to relax looking at the hot springs or volcanoes, or enjoying the wild life on Myvatn!

Group One, Justin and Karl, have planned to do a lot of cycling. From Reykjavik, they are following the road to near Reykholt, and then hiking up to Husafell and back. They are then following the main road through Blonduos and Akureyri to Namaskard and up to Aldeyafoss. They are then hiking south to Askjo, west to Kerlingafjoll, north-east back to Aldeyafoss, and then cycling back along the main road again.

Group Two, Anouska and Naomi, are cycling to Thingvellir, hiking north-east to Kerlingafjoll, east to Sprengisandur, north-east to Askja, and north to Aldeyafoss. Someone will meet them there with bicycles and they will cycle back along the main road to Reykholt, hike up to Husafell, collect their third rock, return to the road and cycle back to Reykjavik.

Group Three, Kim and Lee, are going by road to Thingvellir, hiking north to Husafell, east to Kerlingafjoll, north-east to Askja and then hiking all the way back along the same route, finishing by road again from Thingvellir.

Fig 4.17 • Lake Myvatn in Iceland

Seeking the Sun

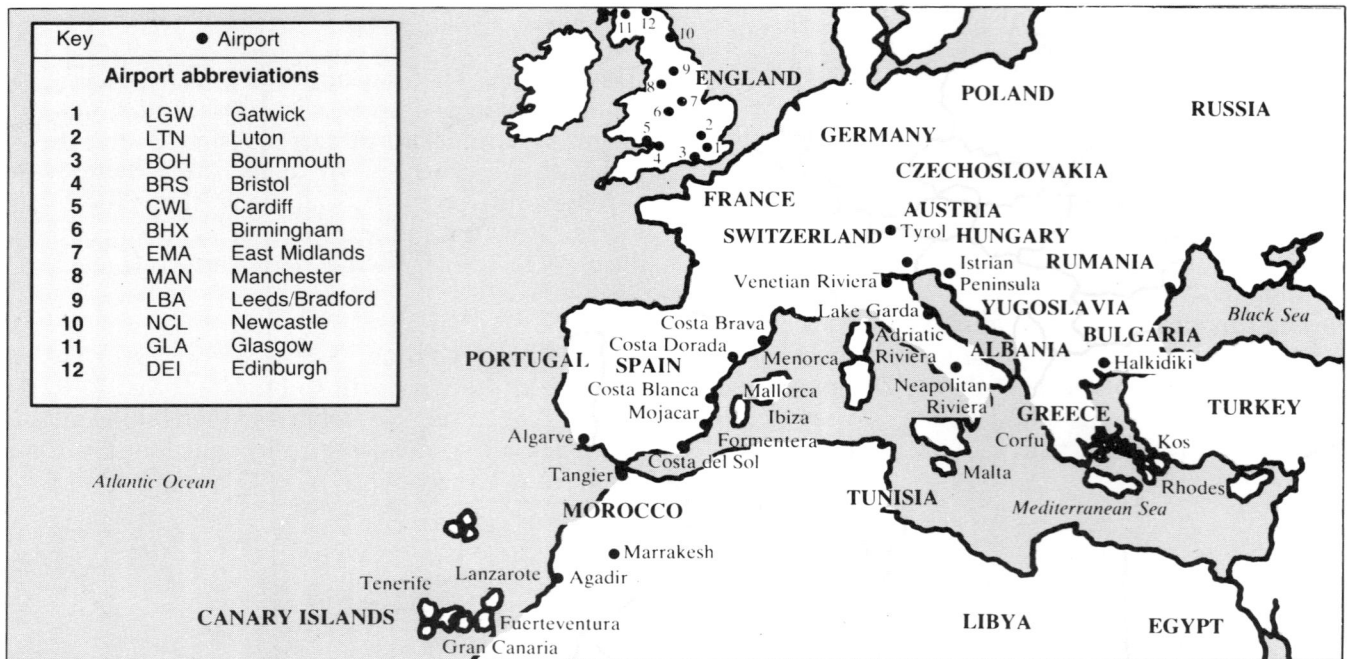

Key		● Airport
Airport abbreviations		
1	LGW	Gatwick
2	LTN	Luton
3	BOH	Bournmouth
4	BRS	Bristol
5	CWL	Cardiff
6	BHX	Birmingham
7	EMA	East Midlands
8	MAN	Manchester
9	LBA	Leeds/Bradford
10	NCL	Newcastle
11	GLA	Glasgow
12	DEI	Edinburgh

Fig 4.18 ● Main airports and destinations of British holidaymakers

The Travel Agent you work for in Morecambe has asked you to use the information given here (from various tour operators) to prepare a big window display. This is to promote sales of winter holidays. The customers include visitors who are staying at Morecambe for a night or two, and those who live in various towns in southern Scotland and north-west England.

Prepare your text and layout for this assignment. Your boss has asked you to emphasise:

- the choice of airports within easy reach of the customers' homes
- the most popular resorts in the Canary Islands, and how far south they are
- the three different Rivieras that are popular in Italy
- the four Greek islands named on the maps

- the five popular resorts in Spain
- that at one resort (which?) you can expect a temperature of over 70°F in January
- that at another resort (which?) you can expect a temperature of 61°F in November
- that at another resort (which?) you can

expect six hours of sunshine a day in November
- the best place for sun throughout the October to April period
- how big the difference is in hours of sunshine in December between London and the Canary Islands
- which of the two airports in Spain is

most favoured by British tourists
- how easy it is to reach Torremolinos and Marbella from Malaga airport
- how easy it is to reach Benidorm, Estoril and Lloret de Mar from their nearest airports
- that Madeira is not really as big as it looks in Fig. 4.20.

You are told you must also make a display panel to show that some visitors in each resort will have arrived by road or by sea. Explain that the pie charts showing arrivals at the airports may not give a full picture of the nationalities of the guests in the hotels. Which resorts would be likely to have visitors arriving by road or sea? From which countries might they have come?

Fig 4.19 ● What's the weather like?

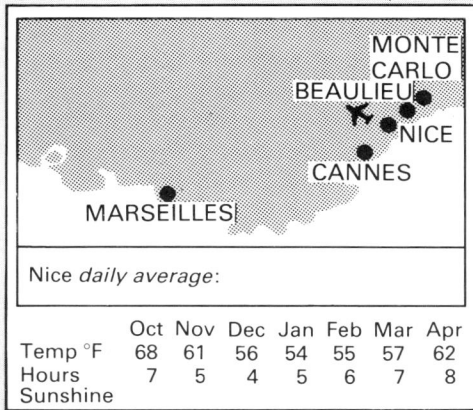

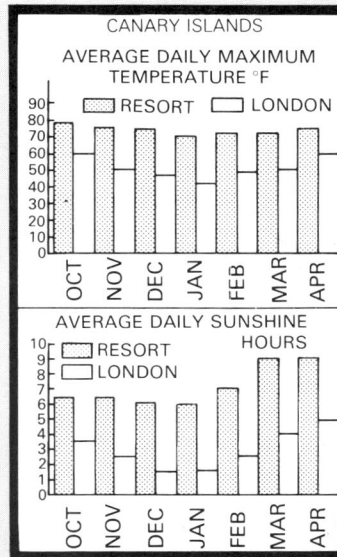

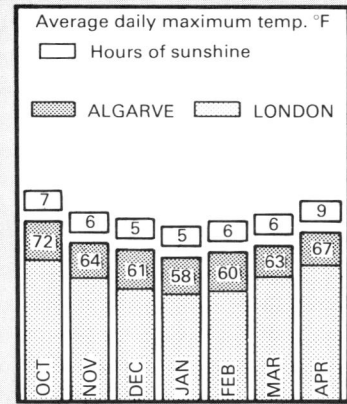

MONTE CARLO
BEAULIEU
NICE
CANNES
MARSEILLES

Nice *daily average*:

	Oct	Nov	Dec	Jan	Feb	Mar	Apr
Temp °F	68	61	56	54	55	57	62
Hours Sunshine	7	5	4	5	6	7	8

CANARY ISLANDS
AVERAGE DAILY MAXIMUM TEMPERATURE °F
RESORT   LONDON

AVERAGE DAILY SUNSHINE HOURS
RESORT
LONDON

Average daily maximum temp. °F
Hours of sunshine
ALGARVE   LONDON

	OCT	NOV	DEC	JAN	FEB	MAR	APR
Hours	7	6	5	5	6	6	9
Temp	72	64	61	58	60	63	67

Fig 4.20 ● Some favourable holiday areas

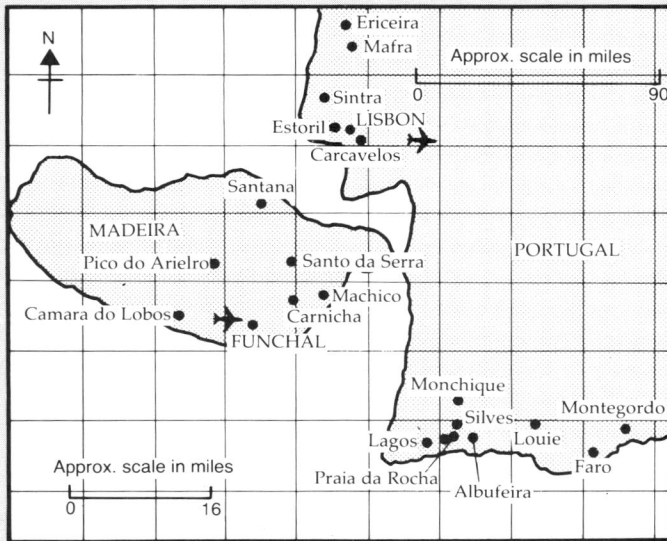

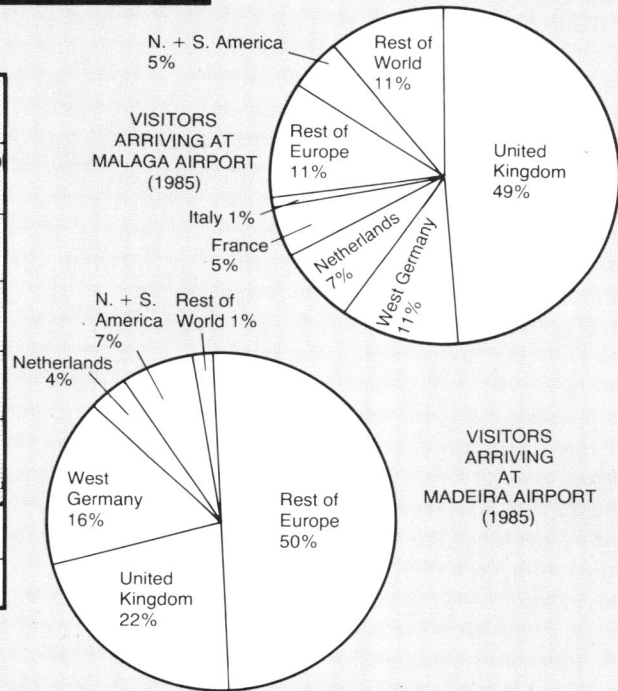

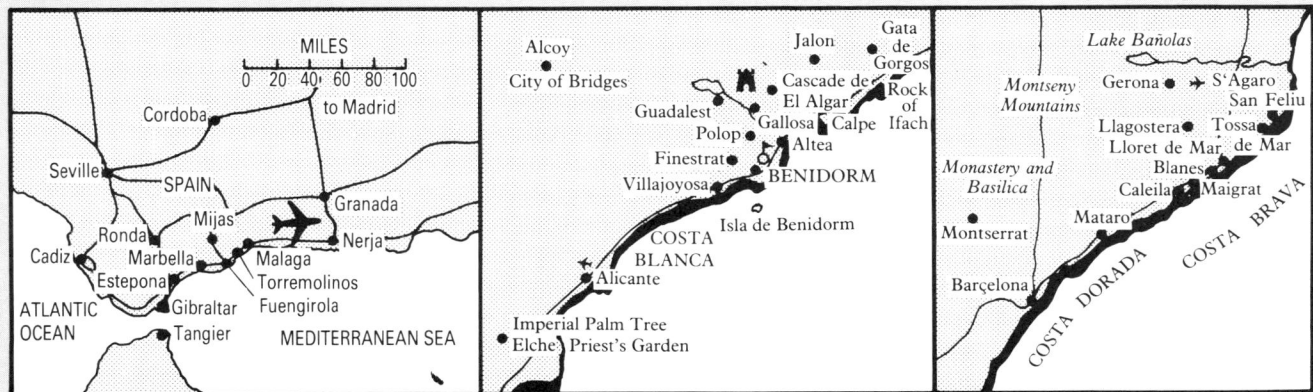

N
Ericeira
Mafra
Approx. scale in miles
0   90
Sintra
Estoril LISBON
Carcavelos
Santana
MADEIRA
Pico do Arielro
PORTUGAL
Santo da Serra
Camara do Lobos
Machico
Carnicha
FUNCHAL
Monchique
Montegordo
Silves
Louie
Lagos
Praia da Rocha   Faro
Albufeira
Approx. scale in miles
0   16

VISITORS ARRIVING AT MALAGA AIRPORT (1985)
N. + S. America 5%
Rest of World 11%
Rest of Europe 11%
United Kingdom 49%
Italy 1%
France 5%
Netherlands 7%
West Germany 11%

VISITORS ARRIVING AT MADEIRA AIRPORT (1985)
N. + S. America 7%
Rest of World 1%
Netherlands 4%
West Germany 16%
Rest of Europe 50%
United Kingdom 22%

MILES
0 20 40 60 80 100
Cordoba   to Madrid
Seville SPAIN
Ronda Mijas Granada
Cadiz Marbella Nerja
Estepona Malaga
Gibraltar Torremolinos
Fuengirola
ATLANTIC OCEAN Tangier MEDITERRANEAN SEA

Alcoy
City of Bridges
Jalon
Gata de Gorgos
Cascade de El Algar
Rock of Ifach
Guadalest Gallosa Calpe
Polop
Finestrat Altea
Villajoyosa BENIDORM
Isla de Benidorm
COSTA BLANCA
Alicante
Imperial Palm Tree
Elche Priest's Garden

Lake Bañolas
Montseny Mountains
Gerona S'Agaro
San Feliu
Llagostera Tossa de Mar
Lloret de Mar Blanes
Monastery and Basilica
Caleila Maigrat
Montserrat Mataro
Barçelona
COSTA DORADA COSTA BRAVA

Quiz

Spot the Errors

What is not correct in the following sentences?

1. Among the sights to see in Snowdonia are the Menai Straits, Derwent Water, Snowdon and the Llanberis Lake Railway.
2. You have to cross the sea to go from Stranraer to Larne, from Belfast to Dublin, and from Dun Laoghaire to Holyhead.
3. The water in a bay, an estuary and a weir will always be saltier than in a stream.
4. The borders of areas of sun or cloud shown in weather forecast maps are the borders of the standard forecast areas.
5. Goods that were regularly taken long distances by canal were corn, milk, building stone and coal.
6. If you are travelling back to England from the Netherlands, you can take a car ferry from Vlissingen, Hoek van Holland, or Den Haag.
7. The motorway between Reykjavik and Akureyri in Iceland now has dual carriageway all the way.
8. The Panama Canal allows ships to travel from the Gulf of Mexico to the Atlantic Ocean.

Clever Clogs

1. Why is there a main railway line to such a small place as Holyhead?
2. What are the traditional languages of Wales, Ireland and Scotland, and what do they have in common?
3. What sorts of plants are likely to grow in (a) a moorland and (b) a marsh?
4. What is the area you live in called in weather forecasts?
5. What are the highest points in England where canals were built?
6. Which is the longest natural lake in England?
7. Which popular holiday resorts in the Mediterranean are in a Communist country?
8. How can you change temperatures from degrees F (Fahrenheit) to degrees C (Celsius or Centigrade), and what temperatures on each scale do most people find most comfortable?

Picture Search

● In which national park is this mountain and what is special about it?

● How would this route be marked on an OS map?

● In which national park is the museum that keeps this boat? On which lake is it?

84

Pick the Word

1. Which one of these counties would you be likely to pass through on a journey from Bradford to Snowdonia? Powys, Staffordshire, Clwyd, Gwent
2. Which of these is in the Republic of Ireland? The Menai Straits, the Mourne Mountains, the Wicklow Mountains, Scafell Pike
3. Which one of these may not be particularly wet? bridle path, marsh, weir, ford
4. Which of these towns is likely to be the first to get rain coming in from the Atlantic? Plymouth, Brighton, Manchester, Newcastle upon Tyne
5. The canal that goes through Blackburn is the Grand Union, Leeds and Liverpool, Grantham, Macclesfield Canal.
6. Which of these is not in the Lake District? Derwent Water, Windermere, Hornsea Mere, Buttermere.
7. Which of these islands is in the Atlantic Ocean? Rhodes, Ibiza, Lanzarote, Malta
8. Which of these is in Portugal? Algarve, Alicante, Benidorm, Menorca

Crossword

Make a tidy copy of this grid, and then solve the clues. Many of the main words have been used on pages 64–83 of this book.

Clues across:
1. Initials of British Rail (2).
2. These two letters can stand for the Isle of Wight (2).
5. A large country in North America that is a member of the Commonwealth (6).
8. The ocean that one has to cross between Britain and Canada (8).
9. Initials of an electronics company found in the middle of the word 'chemist' (3).
10. Wild open country, perhaps high in the hills (4).
12. One of the small oil producing states in the Arabian Gulf (4).
13. Middle two letters of the word 'consists' (2).
14. Either the line called Cancer or the one called Capricorn is this (6).
18. A large country in north Africa that depends on the Nile for water supplies (5).

Clues down:
1. A large city in the north of England between Leeds and Blackburn (8).
2. A sort of pub or small hotel (3).
3. Egypt is dependent on the supplies of this from the Nile (5).
4. The largest ocean in the world (7).
5. The sort of weather a country has over a period of time (7).
6. Initials of the Automobile Association (2).
7. Not at all bright, like a faint star (3).
11. Not off (2).
13. To rest your legs without lying down (3).
15. Consonants in the word 'argue' (2).
16. Middle two letters of 'boys' (2).
17. Consonants in the word 'pipe' (2).

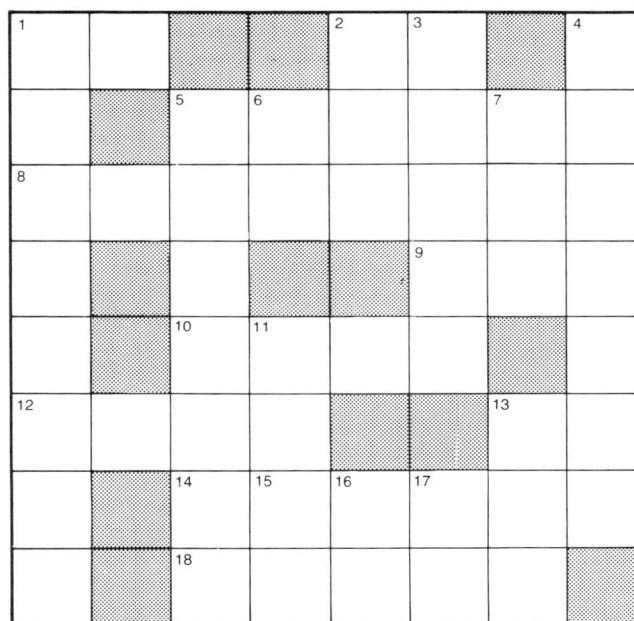

Town Survey Methods

Suddenly you need to visit a town that you have not been to before. You want to

- visit a friend in hospital
- go to a special show or match on your own
- find a workplace for an interview
- collect some goods for your employer
- buy some specialised equipment for yourself
- think of moving there to look for work
- stay there on the way to somewhere else

The work on the next six pages is to help you to practise finding your way around a town that you have not been to before, collecting information you might need.

The questions on pages 88–93 can be used if a visit to either Gloucester or Reading is possible. Otherwise similar questions can be made for a town you do not know that is nearer to your own home.

When you arrive in this town, you should spend between two and three hours (not more) walking around collecting the information that you need. You should then spend another two hours using the information to write detailed reports to answer the questions.

During the time you are exploring the town, it is reasonable to work in small groups (not more than four in a group). But if you are working together, it is essential that everyone does his or her own work. Copying from one another is not allowed.

Take with you a clipboard with a street plan, a copy of the questions, and plenty of paper to make notes. It may be better to use pencils rather than pens out-of-doors. Take two or three with you.

If there are more than a few of you going at the same time, it is important to contact the people in charge of the museums and shopping centres before going.

You should write up the answers to two questions in your report, but it is a good idea to try to collect the answers to three questions as you go around. One of the questions may prove more difficult than you expect. You are not likely to have time to walk around the town more than once.

Fig 5.1 ● Conducting a town survey

It is important to read all the questions you have chosen carefully before you begin your walk. Plan the route you want to take, using the street plan. Make sure you do not walk past places you would then need to visit later!

The notes you make will not be in the same order as the sections of the questions. Take care to put clear headings in your notes. Put down key facts, names, prices, times and other details. You can arrange and explain them all later.

Sometimes you will find that you can buy useful leaflets and postcards, or a copy of the local paper, to help with the details you need. You should never just copy out details from sources of this type. The questions have been designed so that you should be able to find all the information you need without buying anything, without asking anybody, and without going inside any shop or disturbing the local community in any way.

Fig 5.2 ● Kennet lock at Reading

There are many ways of collecting your information:

1. Personal observation – making a sketch, taking exact details
2. Using the Yellow Pages telephone directory in a reference library
3. Using maps set up by the Council
4. Reading displays in windows carefully
5. Making detailed notes of exhibits in the museums, using your own words and including your own feelings about the exhibits
6. Reading timetables displayed at bus and railway stations and looking at maps and information boards displayed there
7. Keeping your eyes open for road signs and other notices
8. Taking note of local activities, traders, delivery vans, big factories and offices

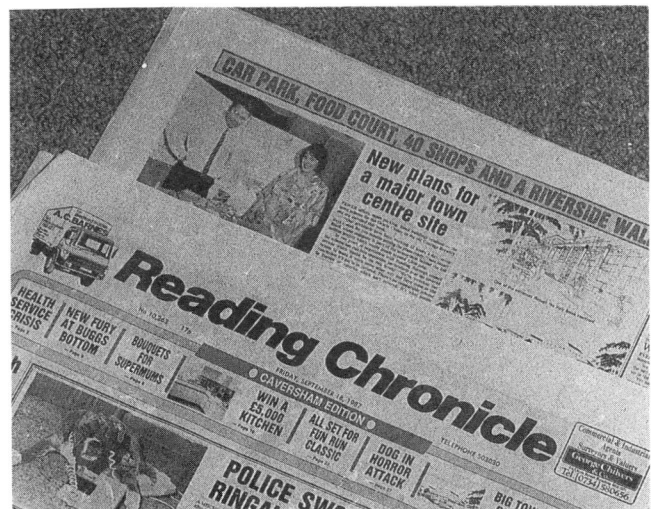

Fig 5.3 ● A local newspaper, useful for information about Reading

Town Survey: Gloucester

You may include leaflets, sketches, postcards or photographs if they will help you to complete your answers. Do not disturb other people in any way. There should not be any need to go into shops. Answer *any two* questions. You have two hours in the city to collect the information you need, and you may have another two hours for writing your answers.

Fig 5.4 ● Gloucester docks

1. *The docks area:*

Part A:

1. In what way is Llanthony bridge somewhat different from most bridges?
2. How do the lock and the weir help to make the docks area a safe place for shipping to enter? What would happen if they were not there?
3. Why is it necessary to have a Customs House at the docks?
4. What were the warehouses built for, and where were the goods in them coming from or going to?
5. There is a canal that leads into the docks at Llanthony bridge. Where does the other end of the canal lead into the River Severn?

Part B:

Choose one of the different areas of the docks (the dry docks, the main basin, the Victoria Dock or the Barge Arm) and write descriptions or draw sketches to show at least six features of design or things provided that make it suitable for its purpose. Explain the **advantages** of each. (Some of these features may be quite small.)

Part C:

Gloucester docks are not as important now as they have been in the past, but people still spend money to keep them going. Explain why the docks continue to be an important part of the life of the City, and the ways in which they can enrich the lives of people there.

2. *Transport by rail, river, and road*

Part A:

1. At what time does the first train leave Gloucester to go to:

● Exeter
● Manchester Piccadilly
● Newcastle upon Tyne?

2. At what time does the first bus leave Gloucester for Swindon on a weekday?
3. What is the number of the bus route you would need to use from the bus station to Coney Hill hospital on the outskirts of Gloucester?

4. What was the difference in track between the 'Cheltenham and Great Western Union Railway' and the 'South Wales Railway'?

5. What are the route numbers of the roads that go from Gloucester to:

● Painswick ● Tewkesbury?

Part B:

1. In which direction is Malvern, and how would you get there by bus? Which company provides the service and when does it run?

2. Where are each of the three places called Woodmancote that are all served by buses out of Gloucester?

3. What routes did the Romans use for transport of goods and soldiers into and out of Gloucester?

Part C:

Explain the advantages and disadvantages of modern Gloucester for making quick and easy journeys to other parts of the United Kingdom or other countries. Comment on things that make it difficult or easy to get from Gloucester to other places, either as a motorist or passenger, or for transporting heavy commercial goods. Consider travel by rail, river, road, sea or air.

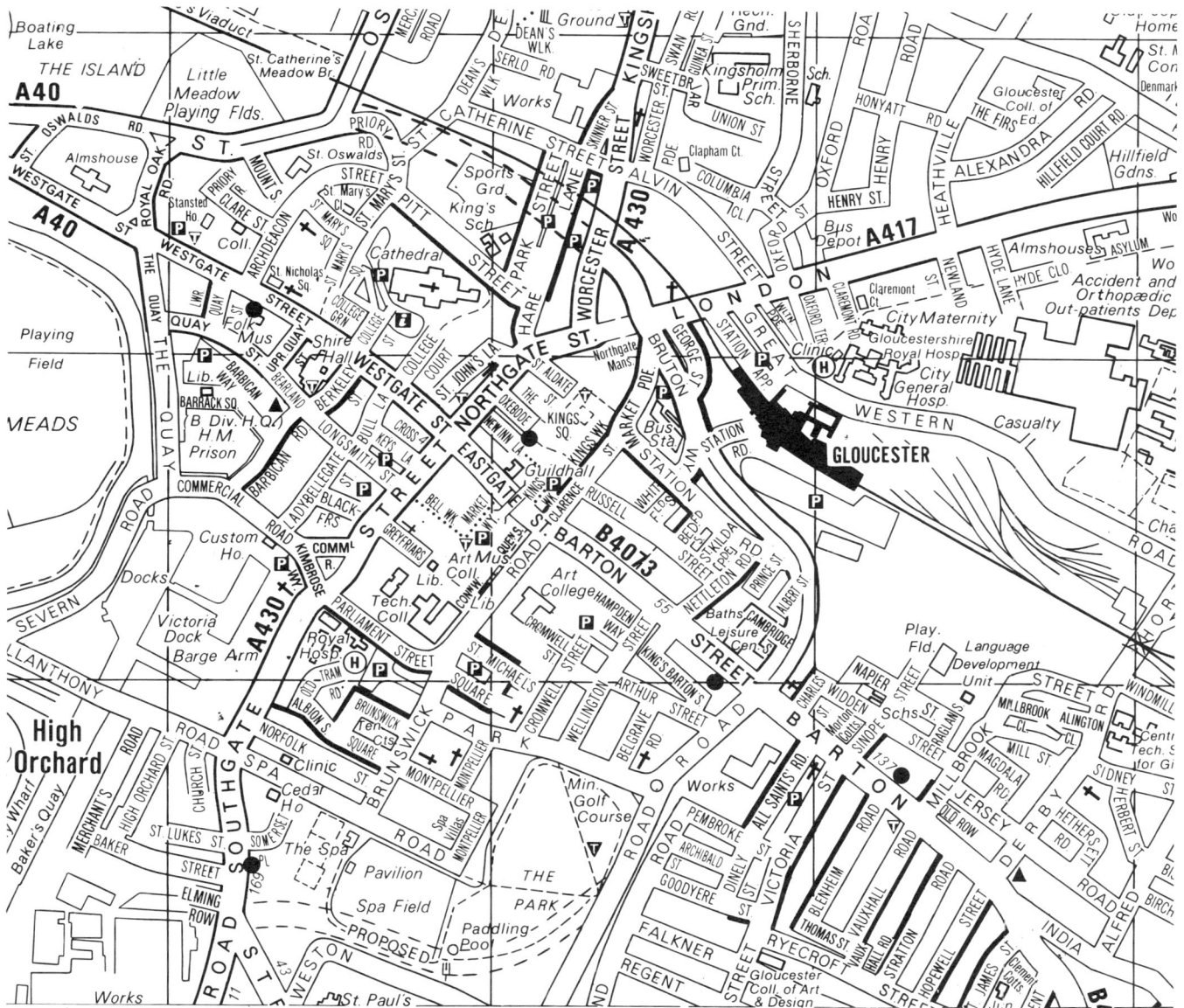

Fig 5.5 ● Part of Gloucester city centre. (*Reproduced by permission of Geographer's A-Z Company Ltd. Based on the Ordnance Survey map with the permission of the Controller of H.M. Stationery Office, Crown copyright reserved*).

3. Landscape and history in and around the city

Part A:

1. What are the old walls that you can see in Eastgate Street outside Boots the Chemist?
2. What old walls can you visit on some days in the summer that are by the King's Square shopping centre?
3. What is the date of St Mary's Gate outside the Cathedral, shown on the notice beside it?
4. Find the main post office. Just inside the door you will find an exhibit. What was the date when the first ship called HMS Gloucester was launched, and how many ships have carried that name?
5. What is the name of the walkway that begins and ends at the Cathedral, and what does it mark?

Part B:

There are a lot of other old buildings in Gloucester. Describe where they are, and how you would reach each of them starting from the crossroads in the city centre.

Part C:

Visit the City Museum in Brunswick Road. In it you will see several displays about the countryside around Gloucester, including the Cotswold Hills, the Severn Vale and the Forest of Dean. Use the information you can collect here to write a description of the area around the city, including the scenery and the wildlife.

4. Shops and town districts

Part A:

1. Visit the Eastgate shopping centre and Eastgate market. One part is called Bell Walk. Why is that a good name for it?
2. What is distinctive about the wall of the British Home Stores in Eastgate Street? Describe some details of it.
3. What are the features of King's Square? What can you buy in that area?
4. Give directions for finding Market Parade from Southgate Street.
5. Why are there more big shops along Eastgate Street than along Barton Street?

Part B:

There are several estate agents with displays of houses for sale in the district. On these displays, as well as on details of buses and elsewhere, you can find out about the different districts in the outer part of the city. Say what these districts are called and what you discover about what they are like.

Part C:

Give details of a tour around the city centre that you would like to make. Give directions in such a way that someone else could follow your route. Explain what you would like to visit, or which shops you would like to go to, and where you would find them.

Fig 5.6 ● Eastgate Shopping Centre, Gloucester

Town Survey: Reading

You may include leaflets, sketches, postcards or photographs if they will help you to complete your answers. Do not disturb other people in any way. There should not be any need to go into shops to answer any of these questions.

Answer *any two* questions. You have two hours in the city to collect the information you need, and you may have another two hours for writing your answers.

1. Reading's links with the world

Part A:

1. How often do buses go from Reading Station to Heathrow Airport during the middle of a weekday?
2. How often do trains go from Reading Station to Gatwick Airport during the middle of a weekday?
3. Which foreign town is twinned with Reading, and in what country is the town?
4. What does the statue of the lion in the Forbury Gardens commemorate, and what places are mentioned?
5. Go to the bus station. What foreign countries are mentioned in the advertisements displayed there? What holidays are offered in them? What countries do you see advertised in the windows of other travel agents in the town?

Part B:
Go to the hallway outside the reference library. Look at the foreign telephone directories. Which English-speaking countries have directories available there?

Go along Union Street (between Friar Street and Broad Street). Name four countries from which imports have been made for sale in these shops, and say what has been imported. Collect this information without going into shops.

Fig 5.7 ● Memorial, Forbury Gardens, Reading

Part C:

Comment on what you have found out about Reading's links with the world. Do you think there are any differences compared to your own town's links with the world? What makes people in a town want to build up closer links with the rest of the world?

2. *Finding out about local history*

Part A:

1. What do the monuments outside the station, outside the Town Hall, and in the Forbury Gardens, commemorate?

2. From the display in the bus station, collect information about when the bridges were built over the rivers in Reading.

3. Describe in some detail what the George Hotel (King Street) is like. Why was it built in this style? What does this tell you about the position of Reading?

Part B:

Visit the ruins of Reading Abbey and also look at the display about it in the museum. Find the Holy Brook and the River Kennet. Describe why this was a good place to build a large abbey in the days before towns had drainage systems and water supplies.

Displays in the museum show other aspects of life for people who have lived in the Reading district in the past. This includes the Romans who lived at Silchester. What evidence is there in the museum that the Romans were importing supplies from other countries?

Part C:

From what you have found out during the day, write a brief account of the development of Reading from the very first settlers up to the present time. Why have Reading's rivers and small hills been important in its development?

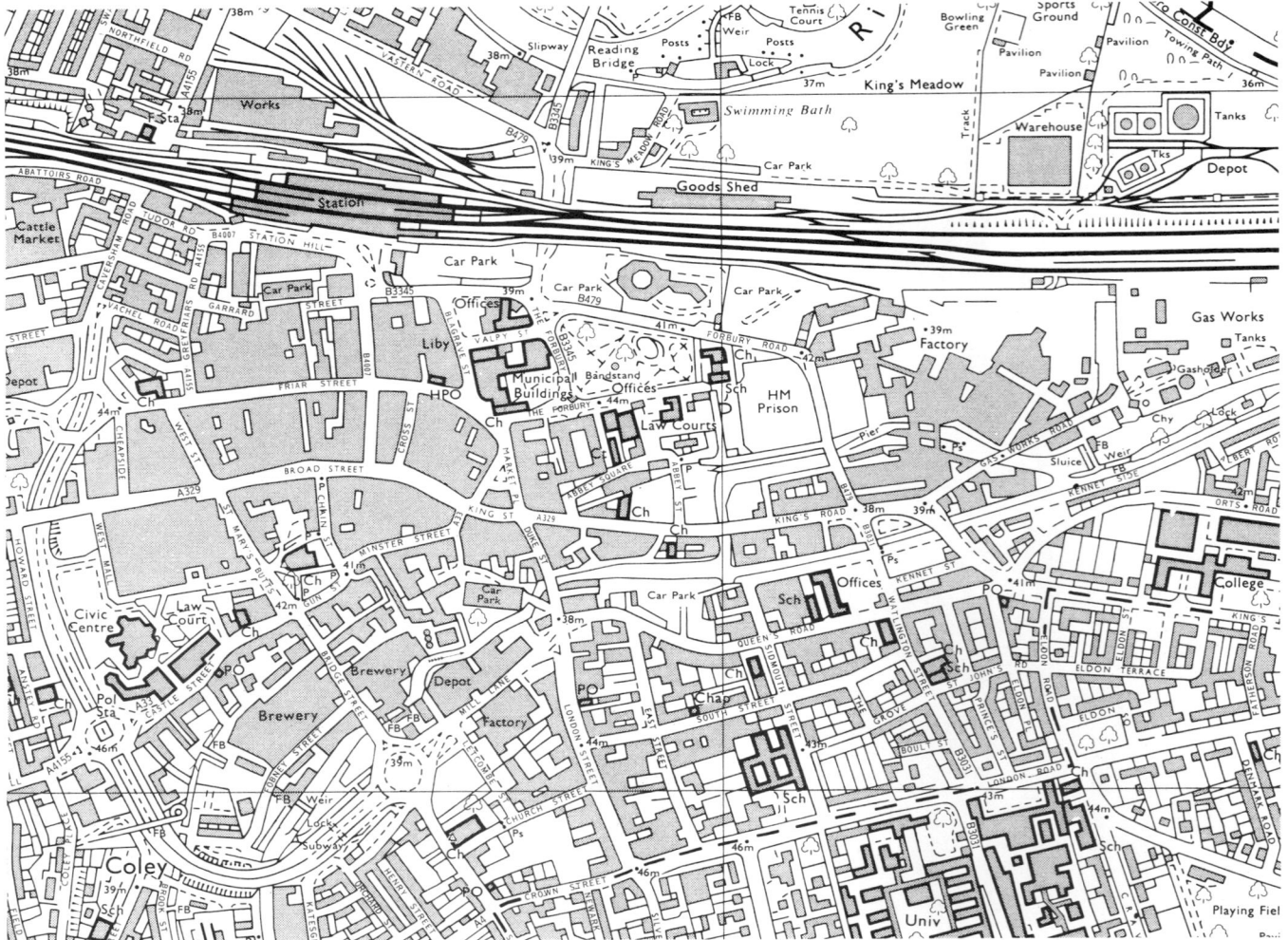

Fig 5.8 ● Part of Reading town centre. (*Reproduced from the 1979 Ordnance Survey 1:10000 map with permission of the Controller of H.M. Stationery Office, Crown copyright reserved.*)

3. Maps

Part A:

1. Go to the steps above the lock on the River Kennet, which you can reach above Willow Street on the Inner Distribution Road. Draw a sketch of the lay-out of the lock to show how boats can pass through.
2. What notices are there for river traffic on the Kennet half way along Fobney Street?
3. Describe the construction of the bridge over the Kennet at Duke Street and say when it was built.
4. Where does the Kennet Side path to Kennet Mouth begin, and how is it marked?
5. The Holy Brook joins the River Kennet just below the Abbey ruins. Why is the Holy Brook not marked on the map?

Part B:

A number of other towns are marked on the signposts in the centre of Reading. What are these towns and which direction are they from Reading? What are the route numbers for the roads to them?

Part C:

If you were moving to the Reading district, and you had enough money to buy eight sheet maps and two books of maps, which sorts of maps would you choose? When would you expect to use them? (Assume that you have no maps of the Reading district or any other sort at present. Comment on the types of maps, and say what districts and what scales you would want to have. You do not need to go into shops to name publishers and prices of particular maps.)

4. Around the town

Part A:

1. Where is the main police station? How do you get to it?
2. Where are the Magistrates' Courts? What features can you see outside the courts?
3. Where are the Assize Courts? Describe the building that they are in.
4. As you go past the gaol, what security systems can you see that make it difficult for prisoners to escape?
5. Describe what the shops are like in Market Way, and the Arcade, both of which are near the Market Place and Town Hall. What sort of business do these shops do?

Part B:

Find places where you could park a car or a motorcycle near the following streets. In each case say what the charges would be for two hours, and give other details of the car park.

- near St Mary's Butts
- near the station
- near Shire Hall
- near Queen's Road

Part C:

Give detailed route instructions for a cyclist to take a short and safe route from the corner of Queen's Road with Watlington Street, to the railway station. Do not go along Forbury Road. Give reasons why you chose your particular route.

Fig 5.9 ● A drawing published in the *Reading Chronicle* showing part of plans for a major town centre development

Index

Note: these references indicate pages where the listing is either named in the text or illustration, or where it is the correct answer to a set question.

Note: these references indicate pages where the listing is either named in the text or illustration, or where it is the correct answer to a question set.

Acknowledgements

The author and publishers thank the following for permission to reproduce material in this book:

Ancient Art and Architecture Collection (3.15), Associated Examining Board (3.1), Automobile Association (1.11, 1.21, 1.24), Aviation Picture Library (1.2, 2.20), Barnaby's Picture Library (1.37), G. I. Barnett and Son Limited (5.8), Janet and Colin Bord (Front cover, 2.18), Brazilian Embassy (3.6), Britain on View (2.4, 2.15, 5.4), BAA Plc (3.9), British Airways (1.41), British Railways Board (1.10, 1.12, 1.40, 4.12), British Telecom (1.44), British Tourist Authority (2.4, 2.13, 2.15, 2.26, 4.3, 5.4), British Waterways Board (4.2), G. A. Cox (3.17), Crown Estate Commissioners (1.28), DFDS Seaways (1.15), Humphrey Dobinson (1.1, 1.5, 1.6, 1.8, 1.17, 1.18, 1.19, 1.20, 1.25, 1.31, 2.1, 2.24, 3.13, 4.9, 5.2, 5.6, 5.7), The Earl of Shelburne (1.30, 1.31, 1.32, 1.33), French Government Tourist Office (1.21), Nance Fyson (1.14, 2.2, 2.7, 3.16, 5.3), Sally and Richard Greenhill (3.7, 4.4), Hallwag Pocket Germany (1.46, 1.47), High Commission of India (3.3), Iceland Air (3.12, 4.17), David Jones (3.5, 3.11), Lindass og Øiseth, Drammen, Norway (1.16), London Regional Transport (1.13), Peter Newark's Western Americana & Historical Pictures (3.10), P & O Ferries, ferry operators to Orkney and Shetland (1.23), Photographic Bureau USPG (3.2), Derek Pratt Photography (4.1), Reading Chronicle (5.9), Royal Automobile Club (1.26), Røyken Ungdomsskole (1.16), Sealink British Ferries Harwich Hook Ferry Line (4.11), Sevenoaks School for Girls (5.1), Spectators Guide to Snowdonia (4.5), The Times (2.29), Thomson Holiday Brochures (4.8, 4.20), Thorpe Park (1.27), 'Today' Newspaper (News UK Ltd) (3.19, 4.7), Viking Publicity (1.7), Bryan Woodfield (1.4).

Back cover: Reproduced from the Ordnance Survey 1:50 000 Landranger map number 172 with the permission of the Controller of H.M. Stationery Office, Crown copyright reserved.